# PROBLEM POSING:
# REFLECTIONS AND APPLICATIONS

# PROBLEM POSING:
# REFLECTIONS AND APPLICATIONS

Edited by
**Stephen I. Brown**
*University at Buffalo*

**Marion I. Walter**
*University of Oregon*

**LEA** LAWRENCE ERLBAUM ASSOCIATES, PUBLISHERS
1993    Hillsdale, New Jersey            Hove and London

Lawrence Erlbaum Associates, Inc., Publishers
365 Broadway
Hillsdale, New Jersey 07642

Lawrence Erlbaum Associates Ltd., Publishers
27 Palmeira Mansions
Church Road
Hove
East Sussex, BN3 2FA
U.K.

**Library of Congress Cataloging-in-Publication Data**

Problem posing : reflections and applications / [edited by] Stephen
  I. Brown, Marion I. Walter.
      p.   cm.
  Includes bibliographical references and index.
  ISBN 0-8058-1065-X (alk. paper)
    1. Problem solving.   2. Mathematics—Study and teaching.
  I. Brown, Stephen I.   II. Walter, Marion I.,
  QA63.P75   1992
  510—dc20                                              92-33746
                                                           CIP

Books published by Lawrence Erlbaum Associates are printed on
acid-free paper, and their bindings are chosen for strength and
durability.

Printed in the United States of America
10  9  8  7  6  5  4  3  2  1

# Contents

# Preface

Since the late 1960's, both of us have been heavily engaged in efforts to place problem posing as a central theme in mathematics education. We have published articles in professional journals, taught graduate and undergraduate courses on the topic at a number of institutions, and have lectured and given workshops at professional meetings in the United States and abroad. Much of our thinking on the topic culminated in our book, *The Art of Problem Posing* — published originally by the Franklin Institute Press in 1983, then by Lawrence Erlbaum Associates with a revised edition published in 1990 (Brown & Walter, 1990).[1]

It was our hope that the book as well as other articles and activities of ours would influence colleagues to (1) further explore the connections between problem posing and solving; (2) further examine the educational values of problem posing; (3) expand upon strategies for problem posing that derive from our schemes, and (4) employ these strategies in situations that we had not previously explored.

Many of our colleagues did in fact take up the challenge. Articles have appeared in professional journals; talks have been given at professional meetings and bits and pieces of problem posing curriculum began to appear in several texts for students.

As we promised in the 1990 edition of *The Art of Problem Posing,* we have created a collection of readings based upon the work of our colleagues in applying and extending the ideas that have occupied a large part of our professional lives for the past quarter of a century. We are grateful to them for having enriched the field so much more than we have dreamed when we first ventured into it.

In addition, we owe much to Hollis Heimbouch, editor at Lawrence Erlbaum Associates, for offering the same confidence, good judgment, and encouragement in the creation and nourishing of this "offspring" that she provided in the "parent" publication. Shawn Vecellio, a doctoral student at The University of Buffalo, created the indexes for this collection.

We are grateful to each of the publishers for having granted us permission to reproduce the articles in this collection. Below is a list of all the essays that appear in this collection together with their original citations. Two articles, the pieces by Feibel and Goldenberg, are published here for the first time. Several of the articles have been edited or excerpted from their original version, and we have occasionally made minor changes in vocabulary for reasons of clarity and in order to modify some sexist language.

# REFLECTIVE ESSAYS

Borasi, R. (1990). The invisible hand operating in math instruction: Students conception and expectations. In T. J. Cooney (Ed.) *Teaching and Learning Mathematics in the 1990's* (pp. 174–182). Reston, VA: National Council of Teachers of Mathematics.

Brown, S. I. (1984). The logic of problem generation: From morality and solving to de-posing and rebellion. *For the Learning of Mathematics*, 4, 1, 9–20. Also in L. Burton (Ed.), *Girls into Maths Go* (1988), London: Holt, Saunders Publishing Co., 196–222.

Brown, S. I. & Walter, M. I. (1988). Problem posing in mathematics education. *Questioning Exchange*, 2, 2, 123–131.

Brown, S. I., & Walter, M. (1990). In the classroom: Student as author and critic. *The Art of Problem Posing* (pp. 119–127). Hillsdale, NJ: Lawrence Erlbaum Associates, Inc.

Buerk, D. (1982). An experience with some able women who avoid math. *For the Learning of Mathematics*, 3, 2, November, 19–24.

Feibel, W. (1977). What if not: A technique for involving and motivating students in a psychology class. Unpublished talk to Western Psychological Association.

Fielker, D. (1983). Removing the shackles of the euclid: 8: "strategies," *Mathematics Teaching*, 103, 48–53.

Goldenberg, E. P. (1990). On the building curriculum materials that foster problem posing. *Seeing Beauty Collections*. Unpublished manuscript of Educational Development Corporation for National Science Foundation.

Jungck, J. R. (1985). A problem posing approach to biology education. *The American Biology Teacher*, 47, 264–266.

Scheffler, I. (1989). Vice into virtue or seven deadly sins of education redeemed. *Teachers College Record*, 91, 2, 177–189.

# ALGEBRA AND ARITHMETIC

Blake, R. N. (1984). 1089: An example of generating problems. *Mathematics Teacher*, 77, 14–19.

**Borasi, R.** (1986). Algebraic explorations of the error $\frac{16}{64} = \frac{1}{4}$. *Mathematics Teacher*, 79, 246–248.

**Brown, S. I.** (1989). How to create problems. In S. I. Brown (Ed.), *Creative Problem Solving*, (pp. 6–8). New York: University of State of New York; The State Education Department.

**Brutlag, D.** (1990). Making your own rules. *Mathematics Teacher*, 83, 608–611.

**Bush, W. S., & Fiala, A.** (1986) Problem stories: A new twist on problem posing. *The Arithmetic Teacher*, 34,4, 6–9.

**Cassidy, C., & Hodgson, B. R.** (1982). Because a door has to be open or closed. *Mathematics Teacher*, 75, 155–158.

**Friedlander, A., & Dreyfus, T. D.** (1991). Is the graph of y = kx straight: *Mathematics Teacher*, 84, 526–531.

**Kissane, B. V.** (1988). Mathematical investigations: Description, rationale, example. *Mathematics Teacher*, 81, 520–528.

**Meyerson, L. N.** (1976). Mathematical Mistakes. *Mathematics Teaching*, 76, 38–40.

**Moses, B. M., Bjork, E., & Goldenberg, E. P.** (1990). Beyond problem solving: Problem posing. In T. J. Cooney (Ed.), *Teaching and Learning Mathematics in the 1990's* (pp. 82–91). Reston, VA: National Council of Teachers of Mathematics.

**Small, M.** (1977). Creating number problems. *Mathematics Teaching*, 81, 42–44.

**Walter, M.** (1989). Curriculum topics through problem posing. *Mathematics Teaching*, 128, 23–25.

**Whitin, D. J.** (1989). Number sense and the importance of asking "why,"? *Arithmetic Teacher*, 36, 6, 26–29.

## GEOMETRY

**Brown, S. I.** (1973). Mathematics and humanistic themes: Sum considerations. *Educational Theory*, 23, 191–214.

**Chazan, D.** (1990). Students' microcomputer-aided exploration in geometry. *Mathematics Teacher*, 83, 628–635.

**Hoehn, L.** (1991). Problem posing in geometry. *Mathematics Teacher*, 84, 10–14.

**Jones, D. L., & Shaw, K. L.** (1988). Reopening the equilateral triangle problem: What happens if . . . *Mathematics Teacher*, 81, 634–638.

**Schmidt, P. J.** (1975). A non-simply connected geo-board: Based on the what-if-not idea. *Mathematics Teacher*, 68, 384–388.

**Sowder, L.** (1986). Looking-back step in problem solving. *Mathematics Teacher*, 79, 511–513.

**Walter, M.** (1987). Generating problems from almost anything. (Parts 1 and 2) *Mathematics Teaching*, 120, 3–7, 121, 2–6.

## NOTE

[1]Different articles in this collection, when originally published, made reference to several different editions of *The Art of Problem Posing*. Because there were changes in page number and in content from one edition to another, we have maintained the original citation of each author. When the editors of this text refer to *The Art of Problem Posing* without qualification however, it is assumed that they refer to the 1990 edition of Lawrence Erlbaum Associates as indicated below.

## REFERENCE

**Brown, S. I., & Walter, M. I.** (1990). *The art of problem posing*. (second edition). Hillsdale, NJ: Lawrence Erlbaum Associates., Inc.

# Introduction

It has become a truism that a problem well posed is half solved. Every field of inquiry, including mathematics, is replete with examples of the influence of problem posing on problem solving. As we and others have pointed out, it frequently takes not only an intellectual *tour de force* but emotional courage as well in order to pose a problem in a way that reconstrues what the culture at large has found acceptable. As we describe in *The Art of Problem Posing* (Brown and Walter, 1990) the history of non-Euclidean geometry is perhaps the most striking example of how the inability to solve a problem over a period of several centuries was a consequence of an entrenched way of posing the problem. Nevertheless, until quite recently, the power of problem posing has not been appreciated by professional educators as an important ingredient in curriculum.

In 1989, for the first time, the centrality of problem posing was recognized as part of a national program for the re-direction of mathematics education. It is included in several sections (especially ones focused upon problem solving) of the visionary publication of the National Council of Teachers of Mathematics entitled, *Curriculum and Evaluation Standards for School Mathematics* (National Council of Teachers of Mathematics, 1989). Furthermore, our work is cited on a number of occasions in the 1990 yearbook of the National Council of Teachers of Mathematics (Cooney, 1990), and is supported in the 1990 publication of the National Research Council, *Reshaping School Mathematics: A Philosophy and Framework for Curriculum* (National Research Council, 1990).

## REFLECTIONS ON THE POWER OF PROBLEM POSING

As you read the enclosed essays, you will notice there are many different ways in which the authors connect up with concepts of problem posing. Some understanding of how our own thinking evolved may not only help place some of the essays, in perspective, but may also enable you to find ways of elaborating upon the perspectives found in each of these essays.

Our earliest and most pervasive theme — superseding the already mentioned insight that problems well posed are more readily solved — is a pedagogical one that derives from an insight of Polya's: Students who pose their own problems are better motivated to solve them than if they are imposed from without — through texts or teachers.

In that context, we wrote extensively about the power of our What-If-Not scheme. The scheme has several levels and it can be applied to many different objects (from selected words in conjectures to situations to definitions to propositions). The fundamental insight behind the scheme is that one can create not only new problems but deeper insights about anything that is *given* by modifying the given.

It is interesting that only after we had written considerably about the concept of modifying or challenging the given did we come to appreciate the power of problem posing that begins by accepting the given. Among the categories addressed are: posing historical and pseudo-historical problems, asking questions that create a reverberation between exact vs. approximate answers, selecting from a handy list of questions of a general nature and others. Such strategies are discussed in detail in Chapter 3 of *The Art of Problem Posing*.

Parallel to the issue of strategies for problem posing, is one that investigates *when* it is that problem posing takes place in relationship to problem solving. It is obvious that problem posing is necessary in order to initiate problem solving, but we have shown how it is that problem posing takes place both during the act of problem solving and afterwards as well. In reading the collection of essays, you may want to attend to where in the context of solving specific problems, problem posing takes place. These are matters that we discuss in greater detail in Chapter 6 of *The Art of Problem Posing*.

Though we hinted at such matters in *The Art of Problem Posing*, in some of our other writing, we have suggested that there exists value of an educational nature in re-examining problems posed by students even when the focus is not upon solving those problems. Students can learn a great deal about themselves when they reflect upon the nature of the problems they tend to pose when given free reign to do so. For example, do they tend to pose easy problems? hard ones? Do they tend to pose problems that isolate topic of mathematics or ones that integrate them? Are they more inclined to select problems with a high degree of visualization or ones that more verbal in quality?

## ON THE AUDIENCE FOR THE BOOK

As in the case of *The Art of Problem Posing*, this collection should appeal to a wide audience—including present and future teachers of mathematics at all levels of education, educators in non-mathematical fields who would like to get a glimpse of how problem posing might be imported into their own areas of interest, college mathematics students, and interested lay persons. Much of the collection does not require more competence than what is acquired in a good high school mathematics program.

Though some of the articles tend to focus upon ideas in the elementary school curriculum, others upon secondary school mathematics and others at the college or university level, we have not so categorized them in the collection. The reason that we have been disinclined to do so is that with minor modification problems and situations *jump* levels. So called simple problems become interesting and more challenging with the slightest What-If-Not variation—as we indicated in *The Art of Problem Posing* and as our colleagues frequently show in this collection. Some of the interpretive text preceding each chapter will suggest appropriate *jumping off* points for approaching the material at higher or lower levels of one's education.

Since we have organized the essays in this collection in such a way that they re-explore many of the themes we developed in *The Art of Problem Posing* and elsewhere, it is not essential that the reader be familiar with these other sources. It perhaps however, would be worthwhile to use the book as a reference guide especially if the essays are being read on one's own.

This collection could be used in an in-service or pre-service course for teachers which focuses upon issues of problem solving and problem posing. In that case, it might be more appropriate to use this collection in conjunction with a systematic reading of *The Art of Problem Posing*. Chapters 1 through 5 in that book encourage the students not only to read about problem posing but to experience that activity as *insiders*. In addition, a teacher of such a course might find it valuable to engage the students in some of the mathematical writing activities elaborated upon in Chapter 7—*In the Classroom: Student as Author and Critic*. These are activities that experienced and pre-service teachers could use with their own students in their secondary school courses, but it is probably a good idea for them to experience it themselves first. We deem the ideas in that section valuable enough so that we have excerpted a few pages from that chapter to start this collection (the first essay in Chapter I, Reflective Essays).

## A NOTE ON READING THE COLLECTION

There is diversity along a number of dimensions in this collection. Some of the essays were published in the States, but a number come from abroad (*Mathematics Teaching* from Great Britain; *For the Learning of Mathematics* from Canada). Some were published originally in mathematics education journals, but others are from sources in the foundations of education such as *Teachers College Record* and *Educational Theory*. While the preponderance of essays focus upon mathematics and mathematics education, some of them point to the relevance of problem posing to fields such as biology or psychology. Some of the mathematical essays will appeal initially to those involved in the elementary school experience while others will draw the attention of those interested in a secondary school or college level audience. In the interpretive text that accompanies each chapter, we shall indicate how ideas expressed for one audience may be re-visited or transformed in order to ready them for a variety of audiences.

While Chapter I reflects upon the centrality of problem posing in general and its educational value in particular, Chapters II and III are clustered around the specific subject matter of Algebra/Arithmetic and Geometry respectively.

Regardless of whether or not this collection is used in a course, there are many ways of reading it. You may wish to delve in at random and read an article or two whose title captures your attention. On the other hand, you may find it worthwhile to at least glance through the interpretive text that precedes each of the three following chapters. Chapter I — Reflective essays — though more general and abstract than the others, does raise issues that might appropriately be connected with many of the specific readings that follow. In suggesting variations and alternatives as well as a rationale for problem posing, a number of the pieces in the first chapter refer back in varying amounts of detail to *The Art of Problem Posing*. It would be a good chapter to read if you find yourself wanting a bit more clarification of problem posing schemes than is offered in a particular article in chapters II and III.

Do keep in mind as you read the essays in chapters II and III that, though we have found it convenient to separate them into ones that have an algebraic or a geometric focus, substantive matters in fact are significantly intertwined and there are good pedagogical reasons to search for the interconnectedness of the two fields. We address this issue again in general terms at the beginning of Chapter III and in particular as we review some of the articles in that chapter.

One penultimate caveat: Do skim through some of the essays as well as the interpretive text in Chapters II and III *before* actually reading the final essays (the *Your Turn* section) in each of these chapters (and *after* bemoaning our pretentious use of words such as *penultimate* and *caveat*). At the end of the interpretive text for chapters II and III we describe the substance of each of these essays in enough detail so that you will have the opportunity to integrate your own experiences before seeing how the authors of *Your Turn* approach a specific problem posing topic. Since these authors did not have the benefit of this entire collection before writing their delightful articles, you all have a significant advantage over them. See what new insights you might bring to bear on the problems posed by these two authors as a result of your advantage. Perhaps Brown and Walter will run a contest for the most creative responses!

## A CAVEAT OF SORTS

It should be clear that our primary interest has been in creating strategies for students to learn the value of and schemes for posing problems on their own. Nevertheless, much of what appears here would be of interest for teachers whose goal is less ambitious. That is, it is possible for teachers themselves to employ these strategies behind the scene, so to speak (without involving the students in doing likewise) for the purpose of *souping up* an otherwise dull set of exercises, problems, situations or activities. Even if students do not engage in posing problems on their own, they could be beneficiaries of a rejuvenated curriculum.

## REFERENCES

Brown, S. I. & Walter, M. I. (1990). *The art of problem posing* (2 ed.). Hillsdale, NJ: Lawrence Erlbaum Associates.

Cooney, T. J. (Ed.). (1990). *Teaching and learning mathematics in the 1990's.* Reston, VA: National Council of Teachers of Mathematics.

National Council of Teachers of Mathematics. (1989). *Curriculum and evaluation standards for school mathematics.* Reston, VA: Author.

National Research Council. (1990). *Reshaping school mathematics: A philosophy and framework for curriculum.* Washington, DC: Author.

# REFLECTIVE ESSAYS:
## Editors' Comments

This chapter focuses upon general issues related to problem posing. It addresses a number of philosophical, psychological and pedagogical questions that are potentially applicable to most of the articles in this collection. As indicated in the introduction, you might find it worthwhile to skim this chapter *before* reading the rest of the collection and to come back to it every so often as well (perhaps after reading the next two chapters). You will find the articles in this chapter helpful not only for the purpose of *applying* the problem posing schemes used in this collection but of *modifying* them and of providing further *rationale* for engaging in the activity as well.

Do keep in mind as you read these pieces that though some of them focus upon issues of the classroom, others are useful for enabling one to better understand new ideas regardless of the context within which they are learned. You might find it enlightening to take a point of view that appears to be directed towards the classroom and see how you might use it in nonteaching situations as well.

We have found it helpful to organize the essays in this chapter into three clusters. The first cluster (including two pieces by Brown and Walter) does in fact have a *pedagogical* focus. The second cluster (by Goldenberg,

1

Fielker, and Feibel) consists of three pieces that *elaborate upon and apply* problem posing schemes found in this collection. The third cluster (by Jungck, Buerk, Borasi, Brown and Scheffler) deals more abstractly with a *rationale* for problem posing.

# Pedagogical Focus:
# The Design of a Course
# Editors' Comments

The first essay, **In the Classroom: Student as Author and Critic** by Brown and Walter, is the only one in this collection that is excerpted from the original book, **The Art of Problem Posing**. Actually, that essay (the first part of Chapter 7 of the book) is an updated version of one of the earliest articles we published together over twenty years ago (see Walter and Brown, 1971). It deals with a theme that, though essentially unappreciated at that time, has become a dominant one in the last couple of years: *writing across the curriculum*. Though it is not always so articulated, a major reason that the theme has gained prominence is that educators have become increasingly aware of the fact that writing is an activity that is not merely the execution of thoughts already well developed, but rather that the activity of writing itself both inspires and evaluates one's ideas.

The scheme of this essay involves the design of a course around the concept of classroom editorial boards. In particular we explore the dual responsibility of students, as both authors and as critics. Though the course is designed for prospective teachers, students of ours have adapted it for teaching in the schools as well as in college mathematics classes.

We discuss three phases of the course—phases that wean the student gradually from that of *receiver-*

*of-problems-to-be-solved* to *creator-of-problems-viewed-from-a-multiplicity-of-perspectives*. Among the perspectives is that of reflector on how one operates as a member of a group in the context of both posing and solving problems.

The second essay, **Problem Posing in Mathematics Education**, also by Brown and Walter, speaks about some important *sensitivities* that teachers need to be aware of as they try to introduce problem posing in the curriculum. We see these sensitivities as pervading the specific phases of the course that we described in the first essay. They are ones that essentially every teacher wanting to make use of a problem posing curriculum would have to take into consideration. We speak of

1. the irresistible pull to view anything that looks like a problem into a problem *solving* mode;
2. the consequent need to find ways of viewing and transforming problems that have educational potential apart from problem solving;
3. three different ways in which problem posing and problem solving are in fact logically connected;
4. the need to be aware of two quite different problem posing orientations;
5. the need to be aware of the social context within which meaning originates and is negotiated in the classroom.

We should say a bit more about (4) and (5) above. It is interesting that many people who have read our work associate problem posing with "What-If-Not," — a mode of thinking that challenges the given. Many of the articles in this collection of readings exhibit that orientation. The second chapter of **The Art of Problem Posing** is devoted entirely to an analysis of problem posing which *accepts* rather than *challenges* the given. Strategies of problem posing, however, associated with accepting the given are summarized in the final paragraph of the portion of our essay that discusses category (4). It is worth keeping in mind however that the two problem posing strategies do feed upon each other in unexpected ways and on some occasions the conceptual line between the two may be blurred. We shall return to this issue in discussing the final cluster of articles in this chapter.

With regard to (5), it is important to be sensitive to the significant tension in each student between the need to create meaning on his/her own and drawing upon the social context of the classroom. We stress that part of the educational agenda ought to include an invitation for the student to reflect upon how he/she is affected by that tension in a very personal way. It is important that strategies associated with co-operative learning, for example, not acquire the status of a goal in the classroom, but rather become an object of inquiry as well. Some suggestions for how that inquiry may be initiated are made in our discussion of phase 3 of the first essay in this cluster.

## REFERENCES

Walter, M. I. & Brown, S. I. (1971). Missing ingredients in teacher training: One remedy. *American Mathematical Monthly, 78,* 399–404.

# 1 In the Classroom: Student as Author and Critic

Stephen I. Brown
Marion I. Walter

How might one teach a course that makes use of the problem generating ideas we have suggested in our writing? There are certainly many possibilities, but we would like to suggest the bare outline of one scheme we developed over a period of several years. Though there are several features of our course, the central conception is that of *student as author and as editorial board member.* Placing the student in such a role is a radical notion because it assumes a kind of expertise normally reserved for professionals. Such a reversal of role however is consistent with our fundamental notion that students ought to participate actively in their own education and not be mere recipients of knowledge.

Our students, for the most part, are undergraduate or beginning graduate students in the field of mathematics education — although, as we indicated in the introduction, several of them have come from fields as diverse as anthropology, law, and history. Very little in the way of technical knowledge was required, although it was assumed that the students had acquired some degree of appreciation for the nature of mathematical thinking.

## COURSE DESCRIPTION

As a start, we will reproduce a catalogue description of our course:

### Generating and Solving Problems in Mathematics

The main purpose of this course is to provide a context which will counteract an approach to mathematics which is characterized by clear organization of content, clearly posed problems, logical development of definitions, theorems, proofs. We intend instead to provide students with some feeling for mathematics-in-the-making. We will engage in and explore techniques for generating problems, solving problems, providing structure for a mass of disorganized data, reflecting on the processes used in the above activities, analyzing moments of insight, analyzing "abortive" attempts.

The main structural feature of the course, which provides a focus for other activities, is the creation of several journals—physical entities, each of which is created by groups of students throughout the semester, final drafts of which are produced for all members of the class by the end of the course.

To create the journal, the class is divided into several editorial boards (with three to five members on a board). Throughout the semester students write papers which they submit to boards other than their own. Each board offers written criticism to authors and passes judgment on the papers submitted. They decide to accept, reject or require revisions of student papers. After they have had some practice in criticizing papers, each board begins to establish a policy indicating what kind of material and what writing style it most admires. Once a policy is established, each board publicizes it so that students can decide to submit to a board that is most sympathetic with their point of view.

Sources for journal articles include:

1. Problems or situations arising out of class discussions.
2. Problems or situations suggested by instructors every so often.
3. Articles on problems appearing in professional journals.

The papers can be a student's first attempt at defining, analyzing, or solving a problem. The students can also extend, solve, analyze, criticize one of the topics previously dealt with in the course. We stress that if a problem is selected as a starting point it is not necessary that it be solved. Papers include discussions of false starts, introspection on insights or misconceptions, and a list of related topics and specific problems generated while solving the original problem.

Not only are attempts (even unsuccessful ones) to solve problems valued, but other activities not strictly related to solutions at all are considered worthwhile. On some occasions, for example, students decide to write about their efforts to understand the significance of a problem. Some even decide to write about what they imagined the history of the problem might have been.

Besides the articles themselves, the journals include:

- An abstract for each accepted article.
- Letters of acceptance (or required revisions) sent to the author. (Sometimes the original draft, a letter requiring revision and the final draft all appear in the journal. They indicate the kind of reflection encouraged among students.)
- A list of interesting problems that come up in class or in small group or editorial board discussions.
- A list of books or articles either related to specific problems that have been explored or that provide general background for topics or articles.

In addition to responding to the instructor's call for papers, each board is encouraged to initiate a call for papers that reflects its own emerging policy. Some boards have requested criticism and evaluation of the course: others have called for additional problem posing strategies beyond those discussed in class; still others have run contests for the most interesting pedagogical or mathematical problem students have experienced.

## ORGANIZATION OF THE COURSE

### Phase 1: Pre-Journal Writing

The style and content of the course change as the term progresses. In the first phase of the course the instructor usually selects topics that are rich as a potential source for solving problems. Though some problem posing is encouraged, the primary focus at the beginning of this phase is on solving problems individually, in small groups, and in a large group discussion.

In order to enable students to become aware of different approaches to problem solving among their peers, we occasionally pair students and have them observe each other's effort at working on a problem. They take notes on strategies used and we discuss the different styles exhibited. We attempt to maintain a descriptive rather than a judgmental tone, for we are not so much trying to teach a "right way" of approaching problems, as we are hoping to make people sensitive to what they actually do. If it does not appear to interfere unduly with their activity, students (especially when they are paired up to listen to each other) think out loud during problem solving in order to aid in a diagnosis of their style of approach. In order to gain a clearer picture of the problem solving strategy used, it is helpful at this stage to give the students problems that require some manipulation of materials rather than pencil and paper alone. The geoboard is a good source of problems for this purpose; so are problems involving objects like toothpicks and discs. The famous Tower-of-Hanoi puzzle (moving discs of different diameters from one spindle to another according to certain rules) is a good one to use. So are ones like the cherry-in-the-glass problem as described below:

> Four toothpicks enclose a cherry. What is the minimum number of picks you can move so that the cherry is outside the "glass"?

## Phase 2: Beginning Journal Activity

After students have begun to become familiar with different approaches to problem solving with their peers, we introduce some readings that (a) describe heuristics of problem solving, (b) distinguish styles of thinking and problem solving, and (c) suggest "blocks" to the activity, as well. We continue to assign readings throughout the rest of the course, but neither in this phase nor in later ones do we have a pre-established set of readings. Although the three categories just described are usually represented, selections are made based on the interest and mood of the students. Many of the readings are selected from the bibliography of *The Art of Problem Posing*.

Among "classics" that we have found useful for such exploration are those by Adams (on blocks to problem solving), Polya (on heuristics), and Ewing (on styles of problem solving). We should stress, however, that these are all popular categories in mathematics education and in psychology as well, and there is a growing body of literature that is both expanding and refining issues in each of these areas. Anyone teaching a course of this sort would most likely receive considerable help by reviewing recent issues of professional journals in mathematics education and psychology, and by consulting with colleagues in related fields as well.

Gradually, we begin to encourage students to pose problems based on the ones they have attempted to solve, but at this stage no explicit problem generating strategies are discussed. At this stage, too, we encourage students to record their attempted solutions, insights and newly generated problems, although no official journal activity is introduced.

After about three or four class sessions, we encourage students to state explicitly some of the problem-posing techniques they have used implicitly in the first phase of the course. At this stage we begin to formalize some of the strategies we have developed in chapters 3 and 4 of the text. During this second stage, students begin the journal activity. We introduce them to new problems as potential starting points for their articles, and also encourage them to return to the problems they worked on during the first phase—this time armed with some explicit strategies for generating new problems from what was perceived to be "milked dry."

Among the criteria we have used to select mathematical topics for the first two phases of the course are the following:

1. All students should have some machinery available to define and attack problems in the area. Some might make use of special cases and diagrams; others might deal more abstractly with the topic.
2. Topics should lend themselves to examination from a number of different perspectives (e.g. algebraic, geometric, number theoretic points of view).
3. Although innocent looking on the surface, topics should have unsuspected depth.
4. Problems should be such that students can be enticed by easily suggested "situations" that require a relatively small amount of formal definition.

What satisfies the criteria just listed depends on the background and sophistication of the students. One could select from an endless number of topics or situations that would both meet the criteria and satisfy the appetites of students ranging from those in elementary school to those doing doctoral work. Many of the topics discussed in *The Art of Problem Posing* have made excellent points of departure for journal writing among our students.

### Phase 3: The Journal in Full Bloom

Once students begin to feel comfortable writing articles and receiving criticism from their peers (usually after the first round) we move into the third phase of the course, in which we select content based on specific interests of students and encourage them to *collaborate* not only in their thinking about problems but in their writing as well. We also have them begin to reflect (in journal articles) on their idiosyncratic styles of thinking. Since students have not done very much in the way of collaboration before, it is helpful to indicate to them how use of different perspectives and reflection upon different experiences might enrich the task of writing up a group paper.

The following is a typical group-written assignment:

Choose a question or observation related to Pythagorean triples and work on it in a small group for a while. For next week each member of the group should focus (in three pages or so) on a different question of the sort indicated below:

1. What did you find out mathematically?
2. What were some of the problem solving strategies that were used?
3. What things that you tried as a group were abortive?
4. How does the group problem solving strategy in this case compare with your problem solving strategies in others?
5. What other problems came up or were created when you worked in this group?
6. What were your emotional reactions? What turned you on? Off?
7. What were the different roles played by people in the group?
8. Other?

We believe it is important for students to reflect on how their styles of thinking affect their ability to work in collaboration with others, and also to see how they perform as a function of who initiates the task they work on. Thus, we encourage students to work at least once in each of the following four conditions, and to reflect in writing (for at least one paper) on the difference in their performance under these varying circumstances.

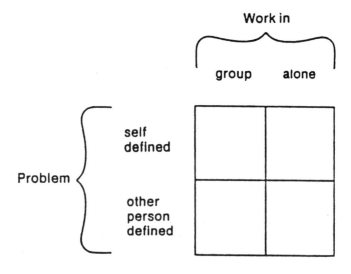

Once the students have reached this phase of the course, they are prepared to put into action some of the board-initiated activity we described earlier (such as a call for papers on topics of their own choosing).

## A WORD OF CAUTION

It is worth stressing that, despite an air of excitement and commitment, we sometimes reach a point of discomfort (usually about one-third of the way through the course or just prior to the requirement that an article be submitted, whichever occurs first) among several of our students. Some students are very concerned about submitting a paper to be evaluated by an editorial board comprised of peers in light of the fact that abortive as well as valuable efforts are disclosed, and in which no final solution of a problem is necessarily expected.

Some students are especially concerned about "airing their dirty linen" in public, especially if they have an image of mathematics as "polished" and "impersonal." We have found it to be both essential and valuable to allow students' discontent to surface and especially to encourage discussion among class members over the issues involved. We have found it worth-while to encourage students to submit an article to one of the class journals in which they try to explore those very issues. The experience of the course is a threatening one, in part because students' prior mathematical experience has taught them to operate in a relatively passive and receptive mode, and to do so non-collaboratively.

It will be necessary, therefore, for teachers who wish to adopt this model to consider different methods of easing students into the role of author and critic. The manner and degree of acclimatization will depend on such factors as age, intellectual sophistication and ability of students to handle peer criticism.

It is important for instructors of such a course to find ways of allaying some of the aforementioned fears that students may have in order to pave the way for significant growth. In order to encourage them to take the kinds of risk we have described, we have found it helpful to assume a grading policy which offers a number of different opportunities to display their many talents. The following is an excerpt of a memo of ours from a recent adventure in teaching a problem posing course to graduate students in mathematics education.

### Grading

Since we want to encourage you to (1) reflect upon your abortive as well as successful efforts in thinking about problems, (2) express your feelings (in writing and orally) about your work on activities associated with the course and (3) accept in a non-threatened way the evaluation of your colleagues, it would be unusual indeed for students who attend the class regularly and who participate in good faith with the class requirements to receive a grade that is not satisfactory.

There will be no examinations or traditional type terms papers submitted at the end of the semester. The grade for each student will depend upon:

1. participation in class discussions and group work.
2. participation as a member of an editorial board (judged by both the process of producing a journal and by the final product).
3. the quality of the articles produced for the editorial boards.

While the instructors will determine that portion of the grade associated with (1) and (2) above, there will be heavy reliance upon the judgment of fellow editorial board members for (3). A student who has pursued the writing of journal papers seriously but has not had a stellar reception by the boards might still receive a good grade based upon performance in the other activities in the course. In addition, though the instructors do not wish to compete with the editorial board judgments during the course of the semester, they will be happy to render independent judgment on the quality of your submitted papers at the end of the semester should you feel that you have not received a fair hearing by your peers.

Regardless of its relationship to grading, however, the role of critic is difficult for many students to assume without first discussing the matter explicitly. Students may wish to discuss both the *value* (and potential pitfalls as well) of criticizing the work of peers, and potential criteria to be used in such criticism. With some encouragement, we find that most students find good reasons for either replacing or supplementing the critical, judgmental, and helpful role normally assumed by the teacher as authority. They come to appreciate that their colleagues may have a refreshingly open and sympathetic reaction to their efforts in approaching new and somewhat risky tasks. In addition, they frequently see their role of critic as one which has considerable potential to be turned "inward" for the purpose of improving their own writing as well.

Once they are persuaded of the value of peer criticism, the editorial boards may need help in coming up with an interesting and coherent editorial board policy. Towards this end, we have found it helpful to have discussions that center on the creation of relevant categories even before positive or negative valences are placed on them. An example of such a category might be "style of exposition." Some boards will eventually select those papers that appear to be tightly argued in a way that may resemble familiar expositions. Others will prize papers that are more chatty in tone.

Other "neutral" categories that students have found helpful for the purpose of beginning to think about the nature of their criticism are: relationship of problem posing to problem solving; creation of new territory versus reflection upon mathematical ideas with which the student has been familiar for a while; "heaviness of tone" (including, for example the place of humor desired in the paper); degree of succinctness. There are many others that instructors and students will come up with in conversations over several weeks, but the important point we wish to stress at this point is that it is helpful to discuss categories at early stages that appear relevant but at the same time are not "preachy" or value-laden.

Each teacher will have to find a way of modifying the rough scheme we are presenting. In fact, we urge teachers who want to make use of our editorial board strategy to do a "What-If-Not" on the scheme itself— depending on the specific circumstances of the students they teach, as well as their particular goals for teaching. We encourage teachers who are using our approach to adopt such an attitude, despite the sense of insecurity that may accompany it, for we believe not only that problem posing and problem solving are important activities for mathematics students, but that the teaching act itself ought to be viewed in a problematic way. In fact, we fear that despite a new interest in problem solving in the mathematics curriculum, many educators will try erroneously to persuade teachers that a particular package or program will guarantee success. We believe that teaching anything ought to be viewed as problematic, and this implies that no topic (especially not problem posing and problem solving) and no teaching approach should be viewed as something to be "bought" on someone else's say-so. In fact, we believe that all of us ought to be plagued regularly by questions like:

- Why is problem solving being pushed as a central curriculum theme?
- Why should it be taught?
- How does problem solving fit in with other things that are important to learn?

An advantage of adopting a problematic and "What-If-Not" attitude towards our proposed scheme is that it may become accessible not only to mathematics teachers who wish to focus on areas other than problem posing, but to teachers of other disciplines as well.

# 2 Problem Posing in Mathematics Education

Stephen I. Brown
Marion I. Walter

In our writings over the past twenty years we have laid out what we consider to be important aspects of problem posing in mathematics and have described various phases that have surfaced as essential components of our accompanying course. We believe that cutting across what we have called phases of the course are a number of *sensitivities* that are or ought to be recurrent themes which should be incorporated into one's formal or informal mathematical experiences if learning mathematics is to be viewed as an act of liberation. In this article we shall attempt to identify these sensitivities. Our focus here will be more upon the pedagogical issues than upon the mathematical content. The reader interested in experiencing the mathematics in greater detail will be directed to appropriate passages in our book *The Art of Problem Posing* (Brown and Walter, 1983) (abbreviated *APP*).

## SENSITIVITY ONE: AN IRRESISTIBLE PROBLEM SOLVING DRIVE

Even before professional mathematics educators ushered in the 1980s as the problem-solving decade, students tended to view mathematics as a problem-solving activity. We would like to tell a poignant anecdote that emerged from our first team teaching experience – an anecdote that perhaps is more responsible than anything else for leading us to identify problem posing as an essential missing ingredient of learning mathematics.

The course we were teaching at the time was primarily focused upon problem solving. In an early stage of the course, we were interested in having the students investigate properties involving triples of numbers known as Pythagorean Triples, the most famous of which is 3, 4, 5. These sets of triples are whole numbers with a very special property. From a geometric point of view, when a triangle is drawn whose three sides have the lengths of an associated triple (as is the case with 3, 4, 5) it turns out that the triangle must have a 90° angle (a right angle). Algebraically, such triples (x, y, z) are whole numbers that satisfy the equation:

$$x^2 + y^2 = z^2$$
(thus $3^2 + 4^2 = 5^2$).

In an effort to involve students in generating such triples in a spirit of inquiry, we presented them with the following question:

$x^2 + y^2 = z^2$. What are some answers?

We received a number of responses to our question. Among them were:

3, 4, 5
5, 12, 13
8, 15, 17
6, 8, 10.

Eventually, a number of students, thinking they would pull our leg came up with answers like:

0, 0, 0
1, 1, 2
− 3, 4, 5.

It was only after class ended, and we began to discuss what had occurred that we realized that the students had not pulled our leg enough! The difficulty was not that there might be even other extreme cases that they had not suggested (like imaginary numbers) but a much more important issue was at stake. The phenomenal thing about that experience was that they were supposedly solving a problem when in fact no problem had been posed! $x^2 + y^2 = z^2$ is itself neither a problem nor a question capable of being answered. In fact a most appropriate response to the request for answers would have been, 'What's the question or the problem?' (See APP, pp. 14, 15).

   None of this is too surprising. It is part of what is involved in developing a common language among people in a particular field of enquiry. By informal convention, $x^2 + y^2 = z^2$ is frequently a signal that we are searching for or dealing with Pythagorean Triples. These are good arguments in defense of the students who had produced the original list of Pythagorean Triples.

Such short-circuiting, however, is problematic for a number of reasons when carried out with people who are first becoming familiar with a body of knowledge, people who do not yet understand the boundaries of a field. Furthermore, a regular diet of such short-circuiting has the effect of curtailing a great deal of enquiry—enquiry that is sometimes experienced more intensely and intelligently among neophytes than those who are deeply entrenched in a field. Even if our goal is one of leading students up well travelled roads, it makes a great deal of sense to encourage them to explore side-paths that may sometimes appear and even be irrelevant as a means of achieving such predetermined goals. (*See* Brown, 1976; 1987 section 1.2.)

Stepping back from the specifics of this experience, we are suggesting the importance of sensitizing students to the nature of their reaction, and to the phenomenon to which they are in fact reacting. If we want students to be in a position to pose their own problems, it is necessary for them to have an understanding of what a problem is in the first place, to be aware of alternative mathematical entities, and to be particularly sensitive to an inclination to view mathematics as exclusively problem solving. While a precise conception of problem is indeed a deep and interesting philosophical problem itself, it is not necessary for students to be in a position to offer a definition of problem in order for them to experience the variety of near and distant relatives.

One such relative is that of a situation. For example, if we present students with a picture of a triangle and ask for an answer or a solution, it would be unlikely that they would respond as in the above example. It is clear that a triangle (in the absence of other information) is a phenomenon or a concept or a situation rather than a problem, though we can imagine some circumstances in which a triangle taken as a whole would be a problem.

## SENSITIVITY TWO: PROBLEMS AND THEIR EDUCATIONAL POTENTIAL

One of the reasons we found students coming up with pseudo-solutions in the case of the $x^2 + y^2 = z^2$ anecdote, was that they thought a problem had been posed. If in fact they had been given a problem, then it surely would have been expected for them to attempt a solution. While such activity is predictable, are there other options? We have barely begun to entertain questions of this sort, and indeed they seem to have a peculiar ring to them. Given a problem, what else is there to do other than to come up with a solution?

Here we have a nice illustration of the confusion of logic and pedagogy. It may indeed be the case that the concept of problem would make no sense in the absence of the concept of solution. If that is the case, does it follow that there are no other acts of educational value to be performed with a problem than to solve it?

One such possibility is suggested by an alternative to the dominant response we received in the case of our anecdote. Students could be encouraged to take a problem and to create it into a situation rather than a solution, that is to 'neutralize' it in some sense. Let us look at a famous problem which spawned an entirely new branch of mathematics (graph theory) two centuries ago:

> The city of Königsberg (now called Kaliningrad), in East Prussia, is situated on the banks and on two islands of the river Pregel. Parts of the disparate city are connected by seven bridges as depicted below. On Sundays the citizens enjoyed taking a walk throughout the city. Is it possible for them to plan the walk so that they can start and stop from the same spot in such a way that they cross over each bridge exactly once?

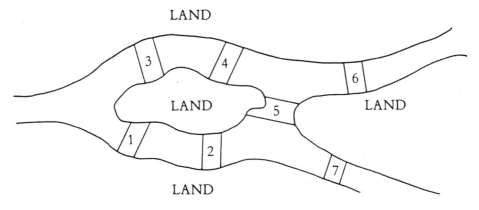

There are of course many ways of making this problem into a situation. One obvious one would be to maintain all of the original description, but to delete the question at the end.

There is much value of an educational nature in so neutralizing problems. As a start, that activity has the potential to make us aware of what precisely is the problem. But more is involved as well. The act of neutralizing a problem to create a situation has the potential to enable one to then create new problems. We thus have achieved a powerful problem posing strategy.

Here is one simple illustration of the creation of a new problem from the famous one above after first neutralizing it.

In how many different ways is it possible to conduct such a stroll (and what should count as a different way)?

It would of course be possible to create a new problem by both modifying and neutralizing the famous one. Below is an example:

Suppose the town consists of the island A and the disc B. Two bridges (1 and 2) connect the two parts of the town as indicated. Is it possible for the people to promenade as the Prussians perchance preferred? [starting from any spot and returning to the same position in such a way that each bridge is crossed exactly once]. (*See* Brown, 1987, Section 5.2 for elaboration.)

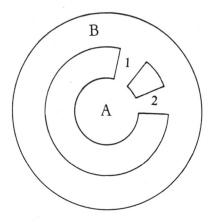

Variations of this sort may not only enable people to understand what the original problem is, but they have the potential to ease the act of 'merely understanding' a purported solution. If one is sensitive to the existence of variations then it makes it easier to understand how reasons apply quite specifically to what is assumed to be given.

Thus, creativity of the sort we have been describing may very well be essential even if one's goal is to understand an already existing body of knowledge rather than something new.

So far, we have spoken of neutralization of problems as a kind of operation that one may perform on them which is clearly different from attempting to solve the problem. Interestingly enough, however, we have still stressed the value of such activity ultimately in relationship to problem solving. Are there ways in which problems may be viewed, transformed, posed that do not connect with an effort to solve them?

There are numerous possibilities, and we have as a profession barely begun to appreciate the options. Given a mathematical problem (or array of problems), what kinds of personal or educational problems could the students be encouraged to pose (and even to investigate) that do not bear directly upon efforts to solve the original problem? We could for example, have students pose problems like: 'Which problems do I find most appealing and why?'. We can ask students to transform given problems into forms they would find more aesthetically to their liking. We can enquire into how it is that students see problems clustering together. We may have them compare the kinds of mathematical problems posed by different students given a particular situation, in an effort to figure out what accounts for the diversity or lack of it.

Perhaps the most significant contribution we can make at this point is to pose an educational problem, one that has not been heard in the mathematics education community: Given a problem, what is there of educational work that students might engage in that is totally unrelated to efforts at *solving* the given problem?

## SENSITIVITY THREE: THE INTERCONNECTEDNESS OF POSING AND SOLVING

Though it surely is possible and on occasion educationally desirable to sever posing from solving, it is also worthwhile to appreciate that there is not one but many ways in which the two activities have the potential to be related. Posing and solving relate to each other as parent to child, child to parent and as siblings as well. That is, though it does not necessarily have to be the case (recall the Pythagorean Triple example), it is usually so that in order to solve a problem, we must have one posed beforehand. But it is also the case that any effort to solve a problem must (by most conceptions of what it means to be a problem) involved a kind of restructuring of the situation which is tantamount to the posing of new problems along the way. Consider, for example, the following:

A fly and a train are 150 km apart. The train travels towards the fly at a rate of 30 km/hr. The fly travels towards the train at a rate of 70 km/hr. After hitting the train, it heads back to its starting point. After reaching the starting point, it once more heads back toward the train until they meet. The process continues. What is the total distance the fly travels?

While there are numerous ways of solving the problem (many of which do not require technical expertise beyond that which is possessed by most sixth graders) it is worth noting that any of the ways that construe this as a problem require that some new problem be posed along the way. As a start, many people who lack the knowledge of infinite series eventually find it helpful to pose a problem that seems to be unrelated to what is called for here. For example, they ask a question like, 'What can we find out about the distance the train travels?' (See *APP,* pp. 118–122 for elaboration of this point). There is yet a third side to the coin which connects problem solving and problem posing, however. Consider the following problem:

Given two equilateral triangles with sides of length a and b, find c, the length of the side of a third equilateral triangle whose area is equal to the sum of the other two.

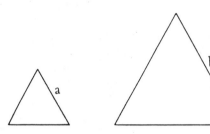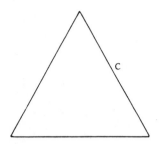

Now, while it always is the case that the solution of any problem has the potential to generate previously unthought of problems, there is something more dramatic exhibited here. In an important sense, one does not fully appreciate what has been solved until one formulates and explores a new question. That is, it turns out, as an apparent quirk of calculation that the length of the third side, c, is such that:

$$c^2 = a^2 + b^2$$

and thus the Pythagorean relationship appears quite unexpectedly. Though we have arrived at a solution to the given problem, we are placed in the interesting position of not being totally aware of what it is we have ostensibly solved until we ask some important questions about the answer. What about this particular problem is leading us to the Pythagorean relationship as a solution? Is it a huge coincidence? (for elaboration, see *APP pp. 107–116*).

Though this phenomenon is not so intimately connected with problem solving as the previous one (as exhibited in the fly/train example) it does highlight the observation that we frequently find connections in places we had not expected. An appreciation for such surprises can in fact be cultivated as we learn to seek them out in places we had not heretofore experienced.

## SENSITIVITY FOUR: COMING UP WITH PROBLEMS

We all come up with problems in many different ways, and for a long time it was accepted as a maxim in philosophy of science that though there may be rational procedures for justifying discoveries once they are arrived at, there is little in the way of heuristics for creating them. There are in fact many different strategies that each of us uses to create problems (if not discoveries *per se*) from other problems or from situations, or even from solutions as we suggested in sensitivity category three. It is an important educational activity not only to teach such heuristics but to find out which ones are already used by our students.

We have found it useful to distinguish two categories of problem generation: accepting the given, and challenging the given. It was these dimensions of problem posing (primarily the latter) that received our earliest and our most enduring attention. In our first articles on the topic (Walter and Brown, 1969; Brown and Walter, 1970), we coined the expression 'What if not?' strategy to describe an elaborate five-step procedure for challenging the given that was helpful in generating problems using different kinds of objects (definitions, organized and disorganized data, theorems, concrete objects, statements) as a starting point. In the 1970 paper, we chose the insight generated by the *faux pas* in which we had asked:

$$x^2 + y^2 = z^2: \text{What are some answers?}$$

We encouraged our students to: 1. make a selection (such as the form $x^2 + y^2 = z^2$); 2. notice the attributes of the object, e.g. it is an equation; or that it is a statement involving three variables; 3. vary the attributes, e.g. suppose we make it into an inequality: $x^2 + y^2 \leq z^2$; 4. ask a question about the new form, e.g. 'Are there any whole numbers for which $x^2 + y^2 \leq z^2$ is a true statement'; 5. analyse the question.

It is interesting in terms of the evolution of the 'What if not?' idea, that it was not until we wrote the second article (Brown and Walter 1970) that we became aware of the fact that we had conflated categories (3) and (4); that is, we had automatically joined the asking of a question with a variation. In fact it is frequently the case that it is a particular question that may suggest a variation, or a variation that may induce what we believe to be a 'natural' question. Once we notice that they have been unintentionally linked, it is possible to create a separation that has the potential to open up worlds we had never imagined.

Though it is possible to become too entangled in the mechanics of the 'What if not?' levels, we do not intend them to be carried out with great abandon, once the strategy is internalized. Intertwining them with some restraint, however, accounts for the most fundamental creative activity. Hofstadter (1982) captures what is involved in activity of this sort when he says:

George Bernard Shaw once wrote . . . 'You see things; and you say "Why?".
But I dream things that never were; and I say "Why not?" ' . . . To 'dream things that never were'—this is not just a poetic phrase but a truth about human nature. . . . How can one dream or even 'see' what is visibly not there? . . . Making variations on a theme is really the crux of creativity. On the face of it the thesis is crazy. How can it possibly be true? Aren't variations

simply derivative notions, never truly original creation? . . . Careful analysis leads one to see that what we choose to call a new theme is itself always some kind of variation, on a deep level of earlier themes (pp. 22–29).

It was several years after we had analyzed and fine-tuned our 'What if not?' strategy, and applied it in our teaching and writing, that we provided similar analysis for what we called *accepting the given*. Given a phenomenon, what kinds of problems can one pose that do not in some sense change what is under enquiry?

Consider for example, the following rather typical curriculum question:

> Given the sequence: 1, 5, 9, 13, 17, 21, 25, 29, 33. . . . What is the fifteenth term?

Why has the curriculum had such a narrow kind of focus? What sorts of questions can students learn to pose that enrich rather than restrict what is given without modifying the given? Consider the following:

What other problems can I pose with that given? (a meta-question?).

What motivation might I have in posing the many different questions to investigate?

Why might people have been drawn to investigate sequences of this type?

The right hand digits for each of the numbers seem to form a cycle: 1, 5, 9, 3, 7. Does that cycle repeat itself? Why?

9 and 25 are both squares. Are there other squares in the sequence? Can I predict what form such squares will have?

3, 5, 7, are prime numbers. Should I expect other prime numbers in the sequence? Why or why not?

These problems are of many different types. Some of them seem to have a general quality about them. They are the kinds of problems one might pose about almost any phenomenon. Others are quite specific in nature in the sense that the meaningfulness of the problem posed requires that one attend closely to the object under investigation. Compare, for example, the first and last questions above. In teaching students to pose problems in an 'accepting the given' mode, it is useful to make that kind of distinction and to search for intermediate ground as well (as in investigating the nature of a pattern, but one that is quite specific to this situation).

Among the general or pervasive categories that we have created with our students are the following: pseudo-history (How might this idea have come about? What could an early form of it look like?); precision modification (How would you describe this phenomenon in rougher, or more precise terms?); internal vs. external enquiry (compare taking the object as a whole in relationship to other objects vs. looking at the workings of the pieces that compose the object). (See *APP* Chapter II for a more detailed analysis.) This kind of investigation has been particularly eye-opening for us and for our students as well because it becomes apparent that one person's change is another's stability.

## SENSITIVITY FIVE: THE SOCIAL CONTEXT OF LEARNING

As we have implied in the above remark and throughout the paper as well, knowledge emerges not in isolation, but in a social context. It is a particularly interesting observation that though the concept of intelligence has been expanded considerably in recent years (Sternberg, 1987 and Gardner, 1983) to include both interpersonal and intrapersonal dimensions, there is a general lack of appreciation for the concept of intelligence as something that resides not in the individual but in society. This alternative view was the hallmark of much of Dewey's thinking. (*See* Dewey, 1935.)

Perhaps one of the most important themes to find expression within a problem posing environment is that of the relationship of the individual to the group. In our teaching we have found it valuable to have students reflect upon how their motivation and cognitive style are influenced by the social context within which problems are posed. How do I pose problems differently when I work alone vs. in a group? Under what circumstances am I more inclined to enquire beyond the original formulation of the problem posed?

We have found it helpful to create formal procedures for group interaction in producing and evaluating mathematical essays — essays in which the students reflect upon the ways in which they pose and solve problems. One such scheme is that of class editorial boards. We usually divide our class up into three or four editorial boards. The student is both author and critic. The members of each board establish editorial board policy, solicit articles from their classmates (sometimes produced individually and sometimes jointly) criticize and evaluate the articles, and finally produce a class journal which includes not only final drafts of accepted articles, but early attempts accompanied by board criticism. (See *APP,* appendix.)

The style of teaching is exhilarating, threatening, mind expanding and places us on an emotional roller-coaster as we all find out and attempt to make public how our mathematical ideas evolve and how they are perceived by and influenced by others. Make no mistake about it; almost everything about the style of interaction runs counter to a deeply ingrained thread of the mathematical culture which places the individual and his creativity on an isolated pedestal. The most clear-cut exemplar of that thread of the culture is R. L. Moore, a mathematician from the University of Texas who was responsible for turning out this generation's leading topologists. (*See*

Moîse, 1965.) It is his role to present students with problems, but that of the students, working in isolation from each other, other faculty members and texts to produce solutions. When a student has a solution, he or she is encouraged to present it to the class. Other classmates may find flaws in what has been produced, but are forbidden from otherwise helping the student for as R. L. Moore claims, no one will know who is ultimately responsible for the solution if it is arrived at as an act of collaboration.

## CONCLUSION

It is possible to incorporate many of the problem-posing types of strategies within the context of standard courses such as algebra, calculus or statistics, courses within which some accepted body of knowledge is taken as a given. Incorporating such strategies as we have described here should have the important instrumental effect not only of motivating and involving the students, but of dissipating a great deal of the fear and anxiety associated with a thoroughly imposed curriculum. It is to be anticipated that to the extent students are given the opportunity to create and explore problems of their own choosing and to reflect upon the educational implications of some of their choices, they will take more responsibility for their own learning. It may take longer to cover predetermined bodies of knowledge, and in fact less material may be covered, but students will most likely uncover a great deal about their style of thinking, attitude towards working with others, and about the purpose of studying and nature of the subject matter as well. In addition, we as teachers may learn that the various talents associated with problem posing and problem solving are distributed in unexpected ways among our students. In the act of co-operation, students and teachers alike may discover ways of valuing such diversity.

In this context, though we have implied a rationale for, and alluded to, important elements of a problem-posing course that we have developed and modified over the years, we find ourselves led to the ironic position of suggesting the dissolution of such a course. That is, it seems to us that these ways of thinking about problem and self are best incorporated in the larger context of curriculum rather than reserved for a precious course here and there. We in fact look forward to the time that a special course on problem posing would be redundant.

It is possible, however to view what we have introduced as 'sensitivity categories' in a more radical way. As students become sensitive to the driving force of problem solving in our present curriculum, and as they and their teachers search for educational uses of problems that move us beyond that valuable but limited conception of what problems are about, we may all learn to acquire the courage to better understand the role of authority in our teaching and learning. We may learn to challenge not only the specific content of a given problem or a given curriculum, but of the ways in which we all interact with and educate each other as well. As teachers we shall perhaps become less interested in locating the best problem-solving curriculum. We perhaps will learn how erroneous the implication is that though mathematical problem solving is valuable for our students because it focuses upon what is unknown rather than known, pedagogical problem solving is a currency that is bought at a cheaper price.

## ACKNOWLEDGMENT

Some of the focus upon educational uses of problems (described in 'Sensitivity Two') was developed by Stephen I. Brown during a leave supported by a fellowship from the Center for Dewey Studies.

## REFERENCES

Brown, S. I. (1976). Discovery and teaching a body of knowledge. *Curriculum Theory Network, 5,* 191–218.

Brown, S. I. (1987). *Student generations.* Arlington, MA: COMAP.

Brown, S. I., & Walter, M. I. (1970). What if not? An elaboration and second illustration. *Mathematics Teaching, 51,* 9–17.

Brown, S. I., & Walter, M. I. (1990). *The art of problem posing* (2 ed.). Hillsdale, NJ: Lawrence Erlbaum Associates.

Dewey, J. (1935). *Liberalism and social action.* NY: Minton, Balch and Company.

Gardner, H. (1983). *Frames of mind: The theory of multiple intelligences.* NY: Basic Books.

Hofstadter, D. (1982). Meta-Mathemagical Themas: Variations on a theme as the essence of imagination. *Scientific American, 247*(4), 20–29.

Moise, E. E. (1965). Activity and motivation in mathematics. *American Mathematical Monthly, 72,* 407–412.

Sternberg, R. J. (1981). *Intelligence, information processing and analogical reasoning: The componential analysis of human abilities.* Hillsdale, NJ: Lawrence Erlbaum Associates.

Walter, M. I. (1987). Mathematics from almost anything. *Mathematics Teaching, 120,* 3–7.

Walter, M. I. & Brown, S. I. (1969). What if not? *Mathematics Teaching, 46,* 38–45.

# Elaborations and Applications of Problem Posing Schemes: Editors' Comments

In the first essay in this cluster, **On Building Curriculum Materials that Foster Problem Posing,** Goldenberg introduces a delightful scheme to accompany a What-If-Not problem posing orientation. It is a strategy that accommodates the fact that students need very different kinds of stimulation at different stages in their inquiry. Goldenberg introduces the concept of "Voices." He incorporates in his writing five voices, each designed to appeal to a different need of the student. He speaks of the voices of narrator, mentor, explorer, philosopher and technical consultant.

The voice of narrator acknowledges the concept of text as telling a story. Story telling is a fundamental part of every culture. We pass on what is unique about members of a group (family, religious group, country) by creating and remembering stories. Some of the stories are known only to a small number of people (as in stories about one's great grandparents and how they came to this country for example); some are deeply ingrained in the culture (as in the apocryphal story of George Washington and the cherry tree).

Story telling is a powerful educational tool. It can be used by teachers to introduce new mathematical concepts to students. Teachers may wish to become the narrators of their own stories and encourage students to do the same. A good source of such stories might be the recollection of events from childhood. Think for example of some of the awe associated with feelings that one has when s/he is flying through space and wondering what it means to believe that there is no end. Think of the fear connected with the fact that one might reach the end and fall off. Think about the awe inspiring realization that between any two points on a rational number line, there must be another point no matter how close together the first two points are (see Borasi and Brown, 1985). We shall explore a specific scheme for using stories in Chapter II in discussing the essay by Bush and Fiala.

The other voices in Goldenberg's essay cajole and encourage the student in a variety of ways. You most likely will find ways of incorporating these different voices not only as an explicit teaching heuristic, but as roles to be taught to your students and assumed by them as they operate together in groups. You may wish to find ways of injecting some of these voices in the articles from the other chapters of this book as well. Such an injection will help suggest new ways of posing problems and new kinds of problems to pose—not only those that are based upon a "What If Not" orientation but those that accept the given as well.

In the second essay, **Removing the Shackles of Euclid: 8 "Strategies,"** Fielker, focusing on the British scene, presents strategies for teachers to create new problems of their own and to encourage their students to do the same. Though his focus is upon Euclidean geometry, his schemes are ones that actually can be generalized to other domains as well. Fielker speaks of *openness, imprecision, limitation, completeness, reversal, depth and variation.*

Though Fielker refers to our work in the context of discussing *variation,* the what-if-not scheme associated with variation may in fact be used to cover some of his other categories as well. Under *openness,* for example, he wisely points out the unnecessary and arbitrary restriction in texts on geometry that focus upon triangles and quadrilaterals. He inquires, "That's all right for quadrilaterals: What happens with pentagons?" While a sensitivity to such *openness* is a sure-fire way of generating these questions, it would also be possible to arrive at the same set of questions by asking What-If-Not on any proposition involving a geometric figure with three or four sides.

You might enjoy figuring out how many of the different strategies he describes can be viewed from the perspective of What-If-Not. It is worth stressing, however that even if a strategy is so reducible from a strickly logical point of view, it is nevertheless valuable to use some of the categories he explicitly identifies as a way of revealing something that may not be perceived when focussing on What-If-Not alone.

It is worth elaborating upon one point of Fielker's in his explicit discussion of What-If-Not under his category of *variation.* His summary of the strategy as one of identifying those aspects of a situation which are variable and then varying them is a good way of holding that category in mind. We stress however, that after a situation is varied, there is usually not *one* but a multitude of questions to impose on the varied situation. The spirit of Fielker's article is such that it would surely support that modification.

The third essay, **What If Not: A Technique for Involving and Motivating Students in Psychology Courses** by Feibel not only suggests some fascinating ways in which the What-If-Not strategy may be applied in another discipline—that of psychology, but it opens up some fascinating pedagogical options. He suggests, for example, that the technique might be used both *before* students have any official introduction to a topic and *afterwards* as well. That is, they might be asked to look at the chapter headings of a text, and by doing a "What If Not" both before and after studying it, come to appreciate what might have been possible but in fact not realizable according to the research in the field. He also explores the value of having students use that strategy in reading classical papers in the field. He comments, "By considering possible directions a field might have taken but for the insights of a paper, students would retain the flexibility of the as-yet-unconverted- and undogmatic." The latter is a lovely thought that has the potential to be applied to many circumstances within which What-If-Not may be employed.

# REFERENCES

Borasi, R. & Brown, S. I. (1985). A *novel* approach to texts. *For the Learning of Mathematics* (Vol. 5, 21–23).

# 3 On Building Curriculum Materials That Foster Problem Posing

E. Paul Goldenberg

Adding the adventure of subjectively original discovery to one's study of a subject is certainly one promising way to get people excited about the subject. In the case of mathematics, it also conveys a central truth about what the subject really is. In the words of Douady (1985), while students "face problems which they believe their teacher can solve," mathematicians "generally face problems nobody knows how to solve." Thus, to experience mathematics as those who love it do, one must spend some reasonable proportion of one's time facing problems for which one does not already have a method mapped out.

This paper deals with the question of how to create the kind of curricular structure and materials that can help teachers provide this kind of experience for students. One approach that seems necessary but that I no longer find sufficient is to base such a curriculum upon a well thought out collection of "good problems" (never mind the definition, for the moment) for students to tackle.

I most strongly sensed the insufficiency of finding good problems for others after *I* had embarked on an exploration because *I* was curious. Even though I had had quite an exciting time of it, I felt seriously let down by the realization that finding such "good problems" for students to do was somehow missing an essential ingredient. What is the fun of discovering what someone has told you how to find? Even the adventures that there might be along the way are hard to experience when you feel that the way has been fully traveled before. And if I *lead* a student to a problem, the student cannot easily feel that the path was untraveled. How many students are likely to believe that their teacher or their text's author would deliberately *lead* them to problems that they, themselves, had never studied?

On the other hand, failing to lead is not a solution either. If I do not provide *some* direction, most of my students are not likely to go anywhere at all, or they may wander around for a while trying this and doing that, but not have any notable adventures. What I need is a kind of guidance for my students that neither leads them by the nose nor leaves them wandering aimlessly: a way of helping students have an adventure without my knowing in advance exactly what adventure they are likely to have.

Adventure, as I'm using it, is of course a metaphor for the exploration one does on a *worthy* and *subjectively new* problem. I say *worthy* because there is little adventure in an exploration if nothing much is encountered along the way. And I say *subjectively*—subjectively original discovery, subjectively new problem—because it is not historical veracity but the students' perception that is important. The problem need not be one that nobody has ever seen before, but it must also not be one that the students' perceive as an Easter Egg hunt—a search for goodies whose location and nature are already fully known by someone who is standing by just waiting while the hunt goes on.

Experts routinely find new and interesting roads to travel. Part of a scientist's overall expertise is a kind of finely tuned intuition about what are worthwhile experiments to perform and about where adventures may lie. But students lack the knowledge and experience from which such intuition grows. To replace (and aid) that intuition, we may guide students not only to some particularly worthy problems, but also to the kinds of heuristics that help them find and pose their own problems. As it turns out, there *are* such methods that can help even a relative novice choose promising sites for mathematical digging.

One method is the "What-if-not?" technique that Steve Brown and Marion Walter describe in their book *The Art of Problem Posing* (1983) and elsewhere. I would characterize it as a teaching method, though it can certainly be adapted for curriculum development as well. The other idea is (at least subjectively) my own, again strongly influenced by Brown and Walter's teaching (though not, to my recollection, by their writing), and by the inspiringly rambling path Wally Feurzeig and I took in the earliest planning discussions for our book, *Exploring Language with Logo* (1987). This technique is illustrated in the *Exploring Language* book, but is described here for the first time. It is a writing and teaching style I've come to call "Multiple Voices."

Although the What-if-not method is well enough described elsewhere, there is sufficient interaction between it and the multiple voice technique that it makes sense to outline the What-if-not method before proceeding with an exposition of multiple voices.

**What If Not?**

This way of helping students find or pose problems (in mathematics) for themselves depends on starting with some familiar mathematical system, listing its features or elements or attributes, and then systematically denying one feature or another — What-if-notting — to create mathematical systems with which they are *not* familiar. These unfamiliar systems, then, can be investigated, and because they bear some relationship to a "familiar" system, students have at least some beginning strategies for investigating the new systems.

To illustrate the process, I will choose Newton's method for finding roots of functions. We begin by teaching a student (for the moment, never mind how) something (for the moment, never mind what) about Newton's method. (We will deal with the how and what in the Multiple Voices section.)

The goal now is to teach the student how to list these things I call "features" of the process. (I am not sure how to say exactly what I mean by *features,* but the examples below should be clarifying.) A visual or geometrical description of Newton's method might lead us to list features like these:

- Begin with a smooth curve, $f(x)$
- Draw a tangent line ($t_1$) to the curve at any $(x_0, f(x_0))$
- Find the zero ($x_1$) of that tangent line
- Build tangent $t_2$ at $(x_1, f(x_1))$
- Look for limiting and non-converging behaviors

Alternatively, we might begin with an algebraic description based on this formula:

$$x \rightarrow x - \frac{f(x)}{f'(x)}$$

Then, the features that we list would be features of the formula.

- Iterate, replacing a value of $x$ with the value on the right side of the formula
- The formula uses subtraction and division
- It uses $f$ and $f'$
- Coefficients and exponents are all 1
- The replacement process uses the same formula at every iteration

Having listed each of these "features" (and more could have been listed, depending on the "grain size" we choose for our features), we can then systematically deny them one by one. For each such change a new system emerges.

Here are a few examples from the geometric description:

- What if the curve were polygonal or discrete points or fractal? (We would, of course, have to come up with a definition for tangent, but that act, itself, is a lovely adventure.)
- What if we drew secants over a certain arc-length or with intersections separated by some fixed $x$ instead of tangents? Or what if we drew normals instead of tangents? Or what if we drew parabolas or circles of best fit instead of lines of best fit (tangents) locally?
- What if, instead of roots ($x$-intercepts), we took some other feature of the tangent line (or circle or parabola), like its $y$-intercept (or center or focus), as a basis for the next iteration?
- What if, instead of concerning ourselves only with the endpoint of the trajectory of this process (the limiting behavior), we look at the trajectory itself?

And here are some examples from the algebraic description:

- What if we used addition, or multiplication, or both? E.g., $x \rightarrow x + f(x) \cdot f(x)$
- What if we used $f'$ and $f'$ or even $f$ and $f'$?
- What if we explored the behavior of

$$x \rightarrow 2x - \frac{f(x)}{f'(x)} \text{ or } x \rightarrow \sqrt{\left| x - \frac{f(x)}{f'(x)} \right|}$$

- What if, at every step, we doubled the coefficient of the $\frac{f}{f'}$ term in

fbcmula?

Each of these suggests an independent and somewhat unpredictable turn. Some of the explorations might (unexpectedly for students) converge on each other. For example the algebraic exploration of $x \rightarrow x + f(x) \cdot f(x)$ and the geometric exploration of normals are really two views of the same thing? This leads to yet new explorations, like "What geometry could we ascribe to $x \rightarrow x - f \cdot f'$ or $x \rightarrow x + \frac{f}{f'}$?"

## Multiple Voices

A process such as the one described above must begin somewhere and must assume some familiarity on the part of the student with some mathematical system. Especially as we want to be able to attract students who have not thought of themselves as mathematically inclined, we cannot assume much background. Wally Feurzeig and I faced a similar situation in *Exploring Language,* in that we did not want to limit the readership to only those people who already had a background in linguistics or Logo.

Also, in their various descriptions of the What-if-not strategy, Steve Brown and Marion Walter seemed to have the *teacher* as an audience. How might a similar effect be built into a curricular structure?

The approach Wally and I took in the book was to weave five voices into a single fabric, changing from one to another (and, thereby, changing texture and perspective) rather frequently. Four of the voices serve complementary roles. Two, the Philosopher/Kibbitzer and the Narrator, maintain a balance between deliberate divergence and the preservation of some sense of direction. Two others, the Explorer and the Mentor, balance the opportunity to engage in open-ended exploration against the need to learn specific facts and skills that are prerequisite for such creative work. In curriculum materials designed to foster problem-posing, the tension between such opposing voices may help to dispel the all too common feeling that the materials deliver the One Right Way, complete with all the questions and all the answers.

## NARRATOR

The narrator is responsible for keeping something like a story line going through the book. In a book of four other voices, it is important for someone to keep track of the overall course: what territories have been visited, what their significance is, what is coming next, where it all might lead, and how the parts interrelate. This voice defines the "structure" of the course.

## PHILOSOPHER/KIBBITZER

The counterpoint to the Narrator's straight path is the *Philosopher/ Kibbitzer.* This voice's purpose is to allow for commentaries to get "off the subject" by making connections *out* of the current curriculum. Its job is the opposite of the narrator's. The narrator has to keep people from feeling lost on a random walk through nobody-knows-where-anymore-because-we've-strayed-so-far-from-our-purpose. The kibbitzer's job is to point to all of the interesting detours—detours that we will *not* take (in order to stay on something like a path), but that an interested student *could* take (or that others have taken). In *Exploring Language,* these connections were often to psychology, or AI, or art, or philosophy of science, or some branch of linguistics that we were not going to explore. In mathematics, they might well be to art, or science, or engineering, or linguistics, or some other branch of mathematics that we had studied earlier or would take up later.

This voice also serves as an "interested" commentator. It is often great fun when a colleague expresses interest in one's work not just by approval, but by injecting a personal (scientific) association, saying "D'ya know what this reminds me of? . . . ." A text cannot, of course, be reminded of anything by what some student will do years after the text is printed, but well chosen associations to what the student is exploring may provide a semblance of that feeling. Furthermore, such playful associations outside of the current topic models the kind of divergent thinking we want our students to show. In *Exploring Language,* for example, the kibbitzer interrupts a study of the grammatical structure of *Dale yells at Dana* to raise a definitional question: Is the verb *yells* or *yells at*? That is, is *yells at Dana* more like *yells where* or more like *berates whom*? This shifts the discussion from the study of the grammar of one sentence to the exploration of the boundaries of grammatical categories.

In a mathematics class, the analogous shift in level leads one to explore definitions, and therefore lends readily to problem posing. For example, one might quite reasonably interrupt a study of polygons to ask how students *feel* about classifying shapes like these as polygons. The first breaks the stereotype, but clearly fits most definitions. In some ways, the second seems a minor change from the first, but in other ways, it seems to violate the "spirit" of the polygon definition. And what about the third, which differs from the second only slightly? What are the consequences of admitting these as polygons? Can we define a clear interior? An *area*? A perimeter? How do these fit with our ideas about how many diagonals a polygon should have?

## MENTOR

In *Exploring Language,* we sometimes needed students to serve in an apprentice role, to learn particular techniques, or to develop certain skills. We invented what we called *Etudes,* and took the musical analogy seriously. Etudes developed the "finger-skills" we needed our students to acquire so that they could handle more complex projects, but, like the Etudes of Chopin, they were (intended to be) art in themselves. Each had some element of investigation in it, and the results of the investigation were always needed for some purpose, but in *Etudes,* the student was always told exactly how to perform the experiments.

## EXPLORER

The natural follow-up to an etude is either an Exploration or an Excursion (with the Philosopher/Kibbitzer) using the results of the Etude and the skills built up in the Etude. Explorations differ from Etudes in that they are more complex, more open-ended, and do not completely specify a method. They may well include the "good problems" that I lamented at the beginning of the paper—the ones we know are great because we've already done them. But in a mathematics course that takes problem posing seriously, this semi-prescribed kind of exploration must also be supplemented by a larger number of problems in which the course is set by the student after a deliberate What-if-not from, say, an Etude. But notice that the What-if-notting *process* is itself a skill to learn. At least initially, I imagine *it* to be Etude-like, and only later can we assume that students will be able to apply it independently as a learned technique.

## TECHNICAL CONSULTANT

This voice is not, properly speaking, part of the pedagogy in the same sense that the other four voices are, but when one's tool system for exploration is a computer, there are times when one needs to understand aspects of the tool in order to be able to use it more flexibly and creatively in explorations. Periodically (and not too often to be a distraction), the technical consultant explains a procedure, or points out some technical options that could have been changed, and so on. Depending both on the nature of the course and the nature of the tools, some of these technical remarks can be elevated from a kind of necessary overhead to become a genuine contribution to the *substance* of the course. For example, changing parameters or options on software is akin to changing features of a problem (and sometimes has that direct effect).

### Building Curriculum to Support Problem Posing

While teachers and curriculum developers must come up with some novel and interesting problems that their new texts, tools and ways of thinking make possible, *their* best problems are not enough. We must, in our curricula, strive to teach students *how* and entice them *to* make up problems of their own, problems *they* deem interesting enough to research on their own or with the colleagueship of their teachers and classmates. Such an emphasis on helping students learn to *behave like mathematicians* (in at least the sense of exploring and posing problems themselves) is intended to give them a real opportunity to feel — and feel correctly — that they are doing original work.

With careful crafting to maintain coherency, a "Multiple Voiced" curriculum design can support this goal. When it is successful, it not only provides a useful structure for the student, but also sets a tone that teachers can emulate: a flexible model of teachers who can give information when needed, and who can serve as experienced masters of the subject when they need to guide students toward problems that are likely to be interesting or worthy in some way, and yet who can also be genuinely curious and puzzled with the students, a senior colleague engaged in some of the same pursuits.

Ideally, it gives students the comfort of having a teacher (text) that can guide and inform, while providing students the room and fostering in them the inclination to decide which routes to travel and how to navigate them.

## REFERENCES

Brown, Stephen I., & Walter, M. I. (1983). *The Art of Problem Posing.* Hillsdale, NJ: Lawrence Erlbaum

Douady, R. (1985). The interplay between the different settings, tool-object dialectic in the extension of mathematical ability. *Proceedings of the 9th International Conference for the Psychology of Mathematics Education,* Vol. II. State University of Utrecht, The Netherlands.

Goldenberg, E. Paul, and Wallace Feurzeig. (1987). *Exploring Language with Logo.* Cambridge, MA: MIT Press.

# 4 Removing the Shackles of Euclid: 8: Strategies[1]

David S. Fielker

The theme of the 1983 ATM Easter Course was 'creating problems', and during the three days not only was a large number of problems created, but it was evident that everyone could respond to this challenge in a wide variety of circumstances, even the young lady who sat down nervously at the beginning of the first session and said, "It's been a long time since I created anything."

If anything is to be done about the sort of geometry we teach then it is no good waiting for new examination syllabuses or textbooks. These days examination boards, thankfully, tend to follow what the schools are doing, and textbook writers follow close behind. But the sort of activity I am advocating makes the use of textbooks inappropriate. It relies on continual cooperation between teacher and pupils, and the stimulus of materials and discussion, not the programming of the printed word.

True, there is a lot of source material around that makes problems and activities available to children through the teacher, and a sharing of these is valuable to all of us. However, we need more of a rationale if we are to be able to continue where the teacher's book stops, change wherever necessary in the light of the circumstances we find ourselves in, respond to the challenges of children, inspire when interest might wane, and above all find the secret of going on when otherwise it may seem time to stop and start something else.

What may have been missing from the Easter course was very much explicit discussion about the sort of strategies one can adopt in order to create problems. Armed with these, we could approach the classroom more confidently in the first place, be more flexible as things progress, and furthermore gradually encourage our pupils to employ similar strategies in order to create their own problems.

The following list applies particularly to geometry though it will no doubt suit in other branches of mathematics. It is by no means complete, but it attempts to identify, illustrate and clarify some of the strategies implied in the previous articles.

## OPENNESS

I hope I have done enough already to convince that the traditional geometry syllabus, whatever type it is, imposes quite arbitrary and stifling restrictions. Where shapes are discussed such discussion is usually confined to properties of triangles and quadrilaterals in a somewhat haphazard and illogical way: yet there is a wealth of material in consideration of alternative classifications, and the examination of how properties reveal themselves in polygons in general. Most geometry takes place in two dimensions: yet there are various profitable ways in which connections can be made between two dimensions and three, and even between two and one!

Perhaps two key questions are of the kind:

*That's all right for quadrilaterals: what happens with pentagons?*

*That's the situation in two dimensions: what is the equivalent in three?*

But we have an obsession anyway with shapes. There are ways of analysing shapes which could reveal relationships between lines and angles, and such relationships could form starting points for later discussions about the shapes produced. Ideas have been proposed, for instance, about diagonals, or about the relations of parallel and perpendicular. As a further example, it is a familiar activity to draw sets of parallel lines in two directions in order to produce parallelograms.

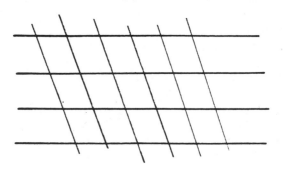

If we adapt the two key questions above, then we can also ask:
*What are the possibilities for sets of parallel lines in three (or more) directions?*

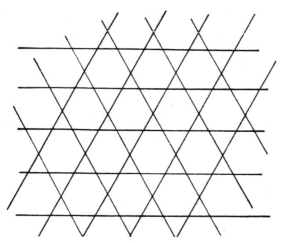

*What are the equivalent situations in three dimensions?*

As has been mentioned before, *Five Sticks* (Fielker, 1975) discusses various ideas concerned with looking at arrangements of lines or line segments. Laurie Buxton has also been looking at the possibilities of such configurations. This arrangement of four lines, for instance,

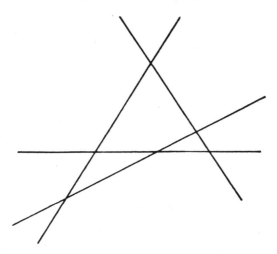

produces one quadrilateral, two triangles, and eight infinite regions bounded by two or by three sides; that should suggest a whole range of problems.

Even lines are not the simplest geometrical entities, and I have played around with some ideas about points, using coloured plastic counters to represent them.

A first sequence of activities begins thus.

*(1) Put down a red counter. Arrange yellow counters so that they are all the same distance away from the red.*

It is fairly simple, but it lends a different sort of dynamic to the definition of a circle. And then someone usually asks, "Can we go into three dimensions?"

*(2) Put down a red and a blue. Arrange yellows so that each yellow is equidistant from the red and the blue.*

Some just put down the yellow halfway between the red and the blue, and some place this one and just two more,

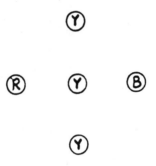

but soon everyone has a line of yellow counters and we can eventually describe it in relation to a line joining the red to the blue.

*(3) Arrange the yellow so that each yellow is twice as far from the red as from the blue.*

This instruction is often misinterpreted and results in a circle centred on the red with radius equal to the distance between the red and the blue! When we sort that out often the yellow line is moved over towards the blue, and the yellows have to be inspected one by one. Then certain key points appear, and eventually something like a closed curve is filled in, with questions about whether it is a circle or an ellipse.

One can then generalise; ask about limits; discuss other loci.

A second sequence of activities begins with three points, which are agreed to form a triangle, except when they are collinear. Now put down four points.

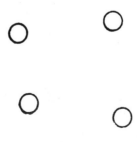

How many triangles can you see?

Some can see four and some eight. We need some rules about when we can 'see' a triangle. We agree on four.

How many now?

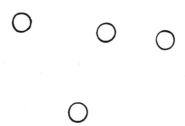

We agree on three. Can you arrange the four points to make two or one or no triangles?

What are the possibilities for five points?

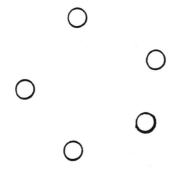

That is not so easy, and it takes a long time to count the maximum number of triangles, with perhaps some colour-coded diagrams.

There are other strategies involved in these activities, but here 'openness' refers specifically to the raw material one is dealing with, the synthetic bits of geometry one looks at: shapes, in a broader sense; lines and points and planes; relations between these; the number of dimensions.

## IMPRECISION

I thought in vain for a better word here; 'vagueness' would have been too strong! What I have in mind is a different kind of openness, a way of putting questions which leaves options free and allows for decisions and choices of interpretation. I cannot better Marion Bird's (1983) example of how she merely asked her class to "make a shape with four squares," and thus produced a wealth of activities based on questions arising from the different ways in which the squares could be joined.

My own oft-used example is *How many different squares can you make on a 25-pin geoboard*? I have to find polite ways of refusing to answer the question, "What do you mean by 'different'?" One must realise, after all, that if one chooses different *positions* as different then one obtains a different answer. After that the only difficulty is that I have to restrain myself while those who begin with the squares with sides parallel to those of the board first realise there are some at 45°, and then that there are other 'oblique' squares. Except that someone usually wants to know if they can count squares like this,

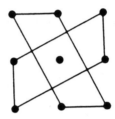

which is a much more interesting problem!

One of the troubles with most texts for children is that the instructions ("tried and tested in the classroom") are absolutely clear and unambiguous — the classroom testing is done to iron out the ambiguities.

A teacher, on the other hand, can present a question in such a way that decisions have to be made. Not only is such decision-making a desirable mathematical ability in itself, but it can lead to a much greater variety of activity.

## LIMITATION

Some problems are too complex to be dealt with all at once, and it can help just to look at certain aspects of them. This is, after all, a perfectly respectable practice among mathematicians, and should therefore be encouraged among children.

Sally, aged 11, was looking at the problem of how many ways three circles can be arranged, a much more complex situation than that for two circles. Somehow (presumably after the two-circle case) this had become related to the number of common tangents. Sally realized that the maximum number of common tangents was 12, so she proceeded to construct cases which gave 11, 10, 9, 8, . . . tangents, a self-imposed restriction that enabled her to solve a sub-problem. As she began to find several possible arrangements of three circles for some fixed numbers of common tangents she began to realize that, in another dimension so to speak, she had a number of sub-problems, and this was a way towards solving the total problem.

Often it is the teacher who has to suggest a limitation, but some limitations seem better than others in terms of the activity they produce. I reported (Fielker, 1974) finding John and Stephen making shapes from tangrams, and inviting them eventually to make all the possible parallelograms from some or all of the pieces. They declined, preferring the simpler limitation of their workcard, which asked for

(i) a rectangle
(ii) a trapezium
(iii) a parallelogram from all seven pieces.

Somehow my limitation had more possibilities, and scope for generality, which the workcard in its complete arbitrariness did not.

In a way all mathematical activities presented to children are limitations, by the teacher, in an attempt to present some manageable part of the curriculum. The Cockcroft Foundation List has a section on Spatial Concepts (about which more in the next article) which includes: *Draw a simple plane figure to given specifications.* I know what this usually means, but in discussing the section with a group of teachers I decided to impose my own limitations on this item, and I asked them to sketch, rather than draw, a dodecagon in which alternate angles were double the size of the angles in between (i.e. the angles were alternately $x$ and $2x$). They found this slightly difficult, very interesting, and generalisable, with the octagon a baffling case! For the generalisations, see the section below on **Variables.** Perhaps it is a characteristic of useful limitations that they are generalisable.

## COMPLETENESS

It has become a favourite strategy to ask *How many*? It is a natural consequence of any sort of classification, as the first article indicated, since whenever one defines a set it seems reasonable to find out how many members there are. But there is more to it than just finding more, because in addition one has to justify in some way that one has found them all, which is really what the article on *Order* was about.

John and Stephen could have found all the parallelograms possible from their tangram pieces, perhaps by considering each size in turn and the number of ways of making each size, thus imposing their own limitation so that they could produce a suitable order, with a consequent proof that all possibilities had been found, a proof built into the construction. In a way they would have done far more mathematics than they did in merely answering their workcard.

Textbooks characteristically avoid *How many*? questions, perhaps because in a formal situation they are difficult to mark. This is certainly true about examination questions, as was indicated in the last article. But any question which asks for one solution or a limited number of solutions *(Draw three nets for a cube)* can always be extended into a *How many*? question.

## REVERSAL

By reversal I mean reversing question and answer, or otherwise putting things the opposite way from usual. It is easy to illustrate numerically by, "The answer is 10; what is the question?"

One might say, "The area is 2; what is the shape?" Frances Purcell (1983) encouraged eleven-year-olds to alter the obvious $1 \times 2$ rectangle by removing pieces and attaching them elsewhere. This produced a lot of activity but it needed limiting in some way in order to become more of a useful problem, and with such limitations the question becomes a *How many*? one. One can be limited to combinations of particular 'half-squares', square dotty paper, a nine-pin geoboard, rectangles. In most cases there are a finite number of shapes which can be found; in the case of the rectangles there is an infinity of solutions which can be suitably described.

A numerical 'answer' can be supplied for other types of 'questions' if one allows measurement to be involved—which on the whole I have avoided. I have never seen the point, for instance, of all the fuss made about 'perimeter' and why it should be a topic at all, let alone one with the same status as 'area'; but given that the idea is a simple one it makes little sense to calculate perimeters of rectilinear shapes, though it does make sense to ask how many different shapes can have a certain perimeter, especially if some limitations can be imposed.

Other reversals are more subtle. I described in *Context* the activity in which pupils were invited to begin with diagonals and so construct quadrilaterals. This is the opposite way round from usual—obviously you cannot have diagonals of polygons before you have polygons. But it makes one think about both in a completely different way, and the classification problems are different.

I have the same feeling working on a microcomputer with Turtle. One nice activity is to construct star polygons. But instead of deciding the order of the polygon, calculating the angle and then constructing it, one can begin with the angle and see which polygon emerges. One soon builds up a feeling for divisors of 360, and perhaps the theory comes later.

In a way some of the tessellation activities suggested in the last article were examples of reversal. Tessellations are usually investigated by putting shapes together, but it sometimes provided more activity if a given tessellation was taken apart, in other words analysed in a different way so that a different method of construction was produced. For instance, the semiregular tessellation of triangles and hexagons, constructed perhaps from considering angles at a vertex, was analysed as triangles (or hexagons) meeting at vertices with hexagonal (or triangular) gaps, and this provided an alternative method of construction for tessellations based on other polygons. Or, as was suggested above, it could be analysed in terms of sets of parallel lines, and this idea could be extended to produce other tessellations.

A lot of geometry is in fact dependent upon the reverse processes of analysis and construction. Each context usually makes one of these more 'natural', and where that is so it is often useful to employ the reverse process in order to create further problems.

## DEPTH

Again I am lost for the right word, and perhaps 'depth of understanding' is what I really mean. This was illustrated best by the discussion of diagonals in *Context,* where there was an invitation to consider the idea of a diagonal in a variety of different contexts, to make decisions about diagonals of polygons in general rather than just quadrilaterals, of reentrant polygons, of crossed polygons, of solids, and to consider diagonal planes as well as diagonal lines. Not only does this provide a deeper understanding of the idea of a diagonal, but it involves a variety of mathematical processes and also produces a wide range of problems.

The Cockcroft Foundation List includes "understand and use terms such as side, diagonal, . . .", but perhaps the Committee did not envisage understanding quite to this extent! They also include "appreciate the properties of parallelism and perpendicularity," and these ideas can receive similar treatment, especially if one includes planes as well as lines.

One way of exploring these two properties is by using those arrow-graphs that used to be called 'Papygraphs', in which, say, letters represent straight lines and arrows represent relations between them.

If the arrows in this graph mean *is parallel to,* what further arrows must be drawn?

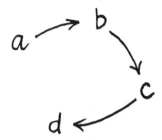

One should also consider whether each line has an arrow to itself, that is whether a line is parallel to itself. This invites a closer inspection of whatever definitions are available, with perhaps alternative choices. *Lines which never meet* needs the addition of coplanarity, and under this definition lines are not parallel to themselves. Can you provide a definition under which they are? Note how the arrow-graph raises the question.

We can use the same diagram for the relation *is perpendicular to,* and again invite insertion of other necessary arrows. Eventually someone asks whether we are in two or in three dimensions, and the two cases need to be considered separately. The contrast is clear, and obviously no line can be perpendicular to itself — not yet, anyway!

Ideas of reflexivity, transitivity and symmetry of relations are implied here, and could be made explicit. Certainly after considering parallelism in this sort of way I never again met the situation I once did, where a class of eleven-year-olds agreed that *a* was parallel to *b,* and *b* was parallel to *c* in this diagram,

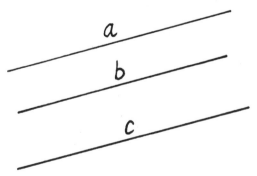

but said *a* was not parallel to *c,* because *b* was "in the way"!

Sometimes it is helpful to discuss what things are not as well as what they are. Most children can tell me that I can draw a circle by fixing one end of a piece of string and tying a pencil to the other. If they forget to tell me that the string must be kept taut then I can soon provoke this by not keeping it taut, or I can ask anyway what happens if I do not. Or I can ask what happens if I use elastic instead of string, or if the string gets wound round the drawing pin attaching it to whatever I am drawing on.

Alternatively I can pin one end of a geostrip (P) and trace a circle through the hole at the other end.

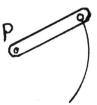

If I use a linked pair of strips, hinged at A, what set of points can I now produce?

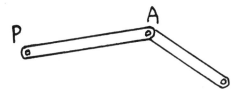

How does it affect the result if I vary the relative lengths of the strips? What happens with three strips? What happens if I use four strips, pin down both ends and trace points at the middle?

All these activities, apart from raising worthwhile problems, teach far more about what circles are and are not than the usual definition, compass construction and 'naming of parts' that is customary in so many courses.

Perhaps I should add that such activities appear to make ideas complex, and I know many teachers who think that their role is to make mathematics easy. In fact all they do is to cover up the complexities, and thereby cause confusion. I believe in facing up to the complexities, even in creating more! The simplicities can come after the struggle.

## VARIABLES

I gave above the problem of constructing a dodecagon so that alternate angles were double the ones in between. In order to generalise this, the strategy is to look at the features of the situation that can be changed, the variables.

So, it does not have to be a dodecagon; I can consider any polygon with an even number of sides. The polygon does not have to be equilateral; I can choose relations between the lengths of sides. The ratio of adjacent angles does not have to be 1:2; I can consider any ratio. It does not have to be pairs of angles; I can for instance choose them in sets of three in proportion 1:2:3.

This is probably the most important and useful strategy available to teachers, and it is one that children can be encouraged to develop. It is equivalent to asking *What happens if . . .* ? or as Marion Walter and Stephen Brown (1969) put it, *What if not*? It is a strategy for turning a single problem into a whole set of problems.

All it involves is identifying those aspects of a problem which are variable, and varying them. Except that it also involves the realization that this is possible and the willingness to do it.

The article on *Order* ended with an invitation to alter the variables in Geoff Giles' problem about making symmetrical shapes by putting these three pieces together.

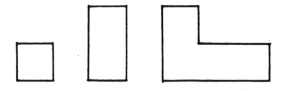

I gave the problem recently to a group of primary teachers, who immediately began to see more ambiguity in it even than I had intended and, after I had refused to give permission or otherwise, made their own decisions about including rotational symmetry, joining halfway along edges, joining at corners, not joining at all, leaving holes and even overlapping!

I then invited them to create variations on the problem. One can of course vary the actual shapes, or the number of shapes, but still base them on squares. They could be based on triangles or hexagons, or rectangles. One can choose from the various rules about how they are to put together. One could change to three dimensions, say by fixing cubes together instead of squares.

In fact, most of the strategies I have listed can be used to develop further problems from this one. The openness is manifest; this is far outside any conventional geometry syllabus. The original problem carries its own imprecision and leaves ample room for interpretation. Limitations are set by any choice of rules. Having done that, each choice implies a *How many*?

question. One can reverse the problem in several ways, say by asking what conditions will give only a certain number of solutions, but the problem is already a reversal since symmetry is usually studied by analysing symmetrical shapes, and here one is asked to construct them. The whole activity gives a deeper understanding of the ideas of symmetry.

## EDITORS' NOTE

[1]The *8* in the title refers not to the number of strategies but to the fact that this is article number 8 in a sequence of articles appearing by the author in the journal *Teaching Mathematics*. The previous seven (with the issue number in parentheses) deal with: classification (#95), interpretation (#96), context (#97), prediction (#98), order (#99), genesis (#101), umbrella (#102).

## REFERENCES

Bird M. (1983). Not necessarily polyominoes. *Mathematics Teaching, 101 & 102,* 12–17.
Fielker, D. S. (1974). Some first year workcards. *Mathematics Teaching, 66,* 24–25.
Fielker, D. S. (1975). Five sticks. *Mathematics Teaching, 72,* 12–16.
Fielker, D. S. (1983). Editorial. *Mathematics Teaching, 102,* 2–3.
Purcell, F. (1983). Editorial. *Mathematics Teaching, 102,* 2–3.
Walter, M. I., & Brown, S. I. (1969). What if not? *Mathematics Teaching, 46,* 38–45.

# "What if Not?": A Technique for Involving and Motivating Students in Psychology Courses

**5**

Werner Feibel

I am sure that most instructors would gladly trade a publication or two for the guarantee of classes filled with motivated and interested students actively interacting with the course material. Well, you can hold on to your publications *for now* because I cannot *guarantee* you such students. But I would like to discuss a technique that has worked quite well in motivating students and in helping them discover the important issues in various course materials. Students have said that this method gives them more confidence in their abilities to organize and deal with information and ideas.

The "What If Not," or WIN, techniques was developed by Drs. Stephen I. Brown and Marion I. Walter while teaching at Harvard (Brown, 1971a, 1976a; Brown & Walter, 1969, 1977; Walter & Brown, 1969) as part of a larger program on problem posing designed for use in mathematic education. Very briefly, the technique works as follows. A problem or topic of interest is analyzed to identify the basic attributes. For example, givens, restrictions or goals are considered for the problems; and assumptions, dimensions, data, variables, etc., for the content areas. This list of attributes is then used to generate new questions or problems by modifying or discarding one or more of the derived attributes. The resulting questions can provide new directions for thinking or research, can simply provide food for thought, *or* can serve as new bases for the acquisition or organization of information about the topic.

Naturally, *not all* the questions will be fruitful. In fact, until the user develops a *sense* for the technique, the dead ends may outnumber the good paths. (Incidentally this possibly discouraging situation can probably be minimized if the instructor initially applies the technique by posing the questions—only gradually getting students to generate the questions for themselves). Nonetheless, the eventually fruitful outcomes should be well worth this initial searching. After all, many of the most dramatic advances in thinking have come about through modifying or dropping firmly held assumptions and preconceptions about a field. e.g. relativity theory, non-Euclidean geometries, behaviorism.

One way of introducing the WIN technique in a classroom situation (though the best way has yet to be determined experimentally) is by taking up the first sets of questions yourself. Present a few examples of WIN at work, and then, introduce the actual steps of the procedure explicitly. Only after this should the students be asked to apply WIN first with your help, and then by themselves. This is the approach taken by Brown and Walter in their math classes. They report encouraging results, though these data are primarily anecdotal rather than the results of rigorous experimentation.

I have used this technique in several ways in psychology classes and discussion groups—generally with quite good results. The most common use has been for particular topics which lent themselves especially easily to such analysis. Again, for the most part, the initial *questions* were posed for the students, who used *their* answers as points of departure for their own questioning.

For example, during the physiological portion of a general psychology course, the coverage on memory utilized among others, the following questions with answers from students. The answers, by the way, were *not* given by the students in technical terms—rather in very intuitive form, though the general *ideas* were there.

I described memory as being physical, in the brain, as involving nerve cells, chemicals, and possibly brain waves. Then I asked "what if memory were not physical"—in the sense of having a structural localization in the brain?

Students suggested that memory might be electrical—e.g., brainwaves, with different patterns possibly being specific memories. These notions are very similar to some of the ideas and work of E. R. John (1973). They also considered chemical storage, and proposed testing this by interfering with the chemical activity of the brain. They were stumped, however, in their efforts to affect only memory by chemical means—perhaps because they lacked the necessary biochemistry background.

After discussing this and mentioning proteins as specific examples of likely chemical stores for memory, I asked the students to consider the possibility that chemicals other than brain proteins were responsible for memory. They mentioned transmitter substances as possible alternatives. It should be mentioned, however, that some of the sophistication in their responses is no doubt attributable to the fact that they had very shortly before this heard a lecture on the nerve and its functioning. Consequently the notions of transmitters and other biochemicals were still quite fresh in their mind. Such responses should therefore be expected. However, this is not necessarily a flaw. Rather, it can be seen as indicating the fact that the WIN technique can be used to follow from, build upon, and solidify previously learned material.

As you can see, a few simple questions led to consideration of many of the major issues in the physiological study of memory.

Similarly, in the section on memory during the cognition and learning part of the course, WIN was used as follows. Again, I am being selective in listing the questions posed, and the answers given by the students.

Memory was described in terms of short and long term components, with STM discussed as limited in capacity and as involving repetition. "What if STM capacity were not limited?"

Students at first could not accept such a suggestion. Then someone suggested that the capacity limitations might exist, but might be involved in some other, non-mnemonic, aspect of information processing. Someone picked up on this and proposed that everything gets in but cannot get out for some reason. This generated great controversy about information overload and similar ideas. People started moving toward a conception of the phenomenon as a perceptual rather than mnemonic limitation—i.e., that only a small number of things is actually seen.

We then discussed possible ways of deciding among the above. Students proposed trying to determine the number of things to which one attends. This led to an objection by others that one would still be faced with the problem of deciding whether it got in and simply could not be retrieved. Finally, someone hit upon the idea of presenting lots of information but only asking for some of it back. With guidance, this was eventually elaborated into something like Sperling's (1960) partial report procedure.

I then asked them to consider what might happen if repetition were not involved in STM. This eventually generated organization as an alternative — which I labelled clustering in the given context. Possible organizations seemed to reduce to categories determined by the situation, or context. Such a notion might be regarded as similar to expectancies in Restle's (1974) memory model.

I then asked the students to consider alternatives to making a distinction between ST and LT memory stores. A continuum model was proposed by some students, with end points such as: specific to general, concrete to abstract, unorganized to organized—the general idea seemingly one of increasing abstraction and organization. With guidance by means of questions about how much was done to information; students came to the view that amount of processing is related to manner or degree of organization. This has certain similarities to Craik and Lockhart's levels of processing model.

Again, a few questions get us a long way. During this process I always try to place the students' intuitive conceptions into perspective.

In a course on the development of cognition currently underway, the topic was prefaced by mentioning the following characteristics of development:

1. it involves change;
2. it has a source;
3. it may or may not proceed smoothly;
4. there are mechanisms of change;
5. there is a course to development.

The possibilities for the attributes were then introduced and students were asked to apply WIN to these, and then to answer the questions. This procedure led to the following possibilities being generated:

The source of development may be internal, external, neither, or both. Whether development is continuous or discontinuous led to a consideration of cumulative versus transformational models of development – though not in those terms.

Similarly, WIN led to the mention of invariant vs. contextual or relative courses of development – again not in technical terms – and to a discussion of particulars to which each would apply. Invariant sequences were felt to apply to general knowledge whereas relative sequences were for specific details and functions.

Students in this course are becoming quite adept at using the WIN technique on their own. They simply use the general chapter section headings as their attributes and proceed from there to a consideration of the readings before actually doing them. They feel this gives them a useful overview and enables them to organize the content in a more useful manner. Furthermore, they claim that the reading goes much more quickly after this, and that it is much more enjoyable.

These examples of the potential fruitfulness of WIN will have to suffice for now. I must forgo examples of WIN applied to specific research or articles – because I would rather mention some of the ways in which WIN can be applied in the teaching of psychology. I am planning a longer version of this talk in which such examples will be considered in more detail.

First of all, WIN need not be limited to use with new students. It also seems ideal for use by advanced students, and *even* researchers, to plot out new lines of investigation and to try out new perspectives on a problem.

Nor need WIN's use be limited to classes of a particular size. WIN *may* seem ideal for seminars or small discussion classes where students could use it to interchange ideas with each other as well as with the course material. However, though possibly not as effective—because of more limited student input—the technique could also be used in lecture classes—especially if supplemented with discussion groups—to provide a basis for the organization of course content.

Certain useful questions could be raised *before* the material is actually covered—to provide an overview like that derived from applying WIN to chapter headings in the development of cognition course. Students should be encouraged to speculate and provide their intuitive answers. These should, *if possible,* be used to demonstrate that *much* of what they will be learning they already realize in basic form.

As the various topics are discussed, WIN could be used by both instructor and students to generate questions and set students thinking about more specific details—alternative interpretations of data, or possible research directions. Modifications of research and theory could also be raised in this way. Such questions would be especially useful if used in discussion sections—since students would then feel the maximal impact of the questions *they* themselves have contributed.

The two sets of questions could be used again—though supplemented by students' own overview and retrospect questions—after the material has been covered. This would enable students to firmly organize and consolidate the information within their more sophisticated perspectives on the material. The vocabulary and way of thinking learned during material coverage would, interestingly enough, provide the framework and basis for its own organization.

The above applications refer to general strategies to be used in *courses.* WIN can also be utilized with particular course *material* in several ways. With respect to the *overall* course matter, students could focus on, and apply WIN to, research or theory. In the former instance, questions would concern variables, Ss, method, internal and external validity, assumptions, interpretations, etc. Theory would be examined through consideration of its concepts, assumptions, predictions, range, precision, supporting data, etc.

Particular studies could also be used as springboards for student participation and learning. Students could be asked to consider the purpose of the study, and the possibilities if the study either had not been done, or had obtained different results. Such speculations often seem very similar to the critiques raised by investigators with opposing theories.

The same could be done with classic papers. By considering possible directions a field might have taken but for the insights of a paper, students would retain the flexibility of the as-yet-unconverted-and-undogmatic. Such flexibility is extremely important in creative learning and thinking. Only by being able to discard strongly held, but possibly incorrect, preconceptions do we manage to make major advances in our under-standing of phenomena.

As I said, all of the above applications are meant for *advanced* as well as beginning students and researchers. In fact, the application of WIN to research and position papers might prove very fruitful to workers who must defend their theories against seemingly devastating and incontrovertible data or arguments.

In addition to being potentially more productive than more usual teaching techniques, WIN is quite enjoyable for students, *and* is highly motivating. Even more importantly, by giving students the opportunity to participate and contribute actively in the course, they will find they are capable of producing and learning a great deal. This *could* increase their confidence in their own abilities, thus leading to enhanced self-esteem, and to an improved self-image.

The nice thing about the WIN technique is its being a structured but simple procedure for sparking student interest and participation. Though many of the questions may lead nowhere, the pearls that reveal a new order to, or insight into, a topic are well worth the false starts.

In summary, then, the advantages of developing such critical and creative dialectic abilities in students should be obvious. In addition to improving command of and insight into the subject matter, the active nature of the process (along with its fruits) should serve to increase students' interest in learning—thus making them more *intrinsically* motivated. This will make school a more pleasant experience, since the student can (at least within fairly broad limits) direct much of his or her own learning. Furthermore, by giving the student such an opportunity for self-direction and participation, the WIN technique can enhance self-esteem and self-concept once the student learns that he or she can deal with the course material.

However, much of this is still speculation. To date, the only data concerning the technique are impressions by Drs. Brown and Walter and myself about WIN's effectiveness in our classes. A program to investigate the value and range of WIN and related procedures is needed.

Whether or not WIN or related techniques prove effective in the long run, the important thing is to continue our search for the best ways of teaching our students.

## REFERENCES

Brown, S. I. (1971a). Rationality, irrationality, and surprise. *Mathematics Teaching, 55,* 13–19.

Brown, S. I. (1976a). From the golden rectangle and Fibonacci to pedagogy and problem solving. *Mathematics Teacher, 69,* 180–188.

Brown, S. I., & Walter, M. I. (1972). The roles of the specific and general cases in problem solving. *Mathematics Teaching, 59,* 5-7.

John, E. R., Bartlett, F., Shimokochi, M., & Kleinman, D. (1973). Neural readout from memory. *Journal of Neurophysiology, 36,* 893-924.

Restle, E. (1974). Critique of pure memory. In R. L. Solso (Ed.). *Theories in cognitive psychology: The loyola symposium.* Potomac, Md: Lawrence Erlbaum Associates.

Sperling, G. (1960). The information available in briefer visual presentations. *Psychological Monographs, 74,* 498.

Walter, M. I., & Brown, S. I. (1969). What if not? *Mathematics Teaching, 46,* 38-45.

Walter, M. I., & Brown, S. I. (1977). Problem posing and problem solving: An illustration of their interdependence. *Mathematics Teaching, 70,* 4-13.

# Rationale: Towards a Multiplistic View of The World Editors' Comments

There is perhaps no other discipline that is so powerful in leaving students with the impression that there is one right way of doing things than mathematics. All of the essays in this cluster provide a glimpse of an alternative pedagogy that applies not only to mathematics but (as Feibel showed earlier) to other fields as well. Each of the essays finds a way of connecting problem posing with that glimpse.

The first essay in this section, **A Problem Posing Approach to Biology Education,** by Jungck suggests how it is that a problem posing approach in the teaching of biology has the potential to counteract an entrenched view of science in general and biology in particular. In many ways this piece draws upon the same kinds of insights that inspired Feibel in the earlier cluster. Jungck, however, draws our attention to some interesting work in the philosophy of science which provides a vocabulary to talk about the limited world view we pass on when we tell our students all about what has already been discovered to be the case in science. In particular, he argues that texts, by passing on only "normal science" (as opposed to revolutionary science, a distinction made popular by Kuhn), do not provide students with a critical view of normal science and also inhibit the student's inclination to ask "questions about matters of concern in their own lives."

Borrowing from some of our work in mathematics education, he argues that it is through the activity of problem posing that both of these deficits — one a view of science and the other a view of self in the world — are rectifiable. He reminds us how ironic it is that biology — a field concerned with life and death matters — has been handled in a way that makes it appear both unproblematic and unrelated to the way people live their lives.

The second and third essays in this cluster, **An Experience With Some Able Women Who Avoid Mathematics** by Buerk and **The Invisible Hand Operating in Mathematics Instruction** by Borasi provide a further theoretical orientation to enable us to understand the power of a problem posing education. Both of them not only draw upon our problem posing conceptions, but upon the research of William Perry, who created a developmental scheme to enable us to talk about different stages one goes through in coming to a mature understanding of what constitutes knowledge.

Buerk outlines the essential characteristics of three clusters of Perry's nine stages: from *dualistic thinking* (everything is right or wrong in some absolute sense) to *relativistic thinking* (everything depends upon context) to a *stage of commitment* (in which one freely chooses an orientation realizing full well its limitations and its tentative status). Buerk has chosen an interesting population on which to investigate possible ways to modify a dualistic view. She has chosen mature women whose view of the non-mathematical world is rather sophisticated. They in fact appreciate the power of relativistic thinking. Their mathematical perceptions however are more limited and they tend to see mathematics in dualistic terms. She speaks about a number of teaching strategies that encourage these women to move to a more sophisticated view of the nature of mathematical "truths" and thinking.

Many of you may find that problems of the sort she introduces and problems of the sort she encourages them to generate are appropriate for moving dualistic mathematical thinkers to more sophisticated views of knowledge. Not only the specific problems, but her teaching strategies may have powerful implications for others. Instead of moving quickly into the activity of problem solving, she encourages them to take the problem itself seriously — by asking questions about the meaning of the problem, by searching for ways of clarifying it and sharing mental images that are conjured up by the problem.

Drawing upon some of the insights of Perry's scheme as well, Borasi further elucidates some of the dysfunctional views of dualistic thinking among high school students she worked with. Some of the dysfunctional views that she describes are the following: You cannot learn from your mistakes, history and philosophy are irrelevant to learning mathematics; spending too much time on a problem is a waste of time.

Borasi discusses a number of strategies that one may use to enable students to acquire a more sophisticated view both of mathematics and of being a learner. As with Buerk, she finds it helpful to draw upon problems that have considerable cognitive conflict – paradoxes or near paradoxes for example. She uses these problems to encourage her students to come to an understanding of what it is they in fact believe about learning and mathematics. In addition, she finds that problem posing activities are helpful in enabling students to come to an understanding of their sometimes fragile but powerful beliefs.

An underlying assumption in both essays is that difficulties students have in learning mathematics may be less related to their inability to understand the specific content and more related to a view of knowledge they hold. Some research connected with these issues especially in relationship to mathematics can be found in Brown, Cooney & Jones (1990).

We have already seen how people have drawn upon research in the philosophy of science and developmental psychology to justify a problem posing orientation in the curriculum. Yet one more rationale for the power of problem posing is suggested by the fourth essay, an excerpt from **The Logic of Problem Generation: From Morality and Solving to De-Posing and Rebellion** by Brown. Drawing upon recent research in moral education as it relates to the development of females, Brown has suggested that a shift from a focus on "the rules of the game" to a desire to "connect with" and to care about people and events that implicate them is called for if we are to understand what it is that many females find lacking in their mathematics education.

It is that perspective that encourages us to elevate a "meta-question" that has been suppressed for a long time in curriculum. That is, we need to find ways of encouraging our students to pose problems like the following to their teachers:

1. Why are you asking me to do this problem?
2. Why are you asking your question of me in the way you do?
3. What do you expect me to get out of this encounter?
4. Is it worth my while to invest a lot of effort in this problem?
5. Is this problem of only "local interest" or might you be asking me to think about it because you see it as connected with a totally new unit or a revised view of what I have been thinking all along?

What is important to understand as you read this essay is that there is potentially something quite radical at stake as we begin to listen to the female voice. It is not only that we need to find ways to enable females to more readily participate in the experience of mathematics (by, for example, encouraging, cajoling, reinforcing them) but that their voice is telling us something important about what has been missing from the mathematics education of all of us.

As you read the essay by Brown, see if you can figure out other ways in which mathematics education in general and problem posing more specifically might be introduced and re-constructed once we take seriously what it is that the female voice is whispering.

The last essay in this cluster, **Vice Into Virtue or Seven Deadly Sins of Education Redeemed** by Scheffler is a nice capstone to our Reflection Chapter and in particular to a cluster focusing upon a multiplistic view of the world. Scheffler's article provides both a mild and a strong support for the activity of "What-If-Notting." In the mild sense, he points out the specific power of what he calls "negative thinking." Much of science progresses not by finding all the evidence we can to support a theory but by searching out evidence that might destroy it.

It is in the act of searching out consequences of negating something small like an assumption or a proposition or something large like a whole theory that we find out the unexpected scope and the potential strength of the assumption, proposition or theory. We spoke in the introduction, for example, about the history of non-Euclidean geometry. For several centuries, people were investigating the "wrong" question because they were incapable of posing the problem in such a way that it might negate what we held as some deep-seated beliefs. That is instead of asking *whether or not* the parallel postulate could be proven from the other postulates of Euclid, most people asked *how* it could be proven.

To the extent that we want mathematical experiences to "transfer" to people's lives, there are important reasons to encourage the act of negation in mathematical thought. That is, most people are inclined to view their friends as people who will support what they believe. In fact, as with the development of science and knowledge more generally, we may learn more from friends who are willing to challenge our cherished beliefs. We thus are enabled to find strong reasons to both test and support what we think is the case.

There is a movement in the schools to infuse much of the curriculum with critical thinking. But as Paul (1990) argues so persuasively, critical thinking is not a matter of learning atomistic techniques but rather of subjecting our deeply held beliefs to scrutiny. A large part of that scrutiny involves considering the opposite of what we believe—its negation.

It is in this sense that Scheffler's essay makes a strong case for What-If-Notting. He not only looks at the act of negation, but he considers six other *vices* of education and he locates a strength in each—vices like procrastination, and forgetting. In each case he considers the negation of a well established educational proposition. What he arrives at, in the spirit of a dialogue, is not the opposite of the original point of view but the modification of it so that we come to know not, for example, that procrastination is good rather than evil, but that there are some important circumstances under which we all ought to procrastinate.

You might enjoy relating the logic of some of the mathematical problem posing done in the essays that follow to the educational matters of the sort Scheffler explores in his essay. It might be valuable for you to see what it is that the two activities have in common. What kind of What-If-Notting done in the essays that follow is like the What-If-Notting that leads us to an understanding that *idleness* (for example) has its advantages? What are some other well entrenched vices that you and your students have accepted (about homework, tests, what good teachers always do), that might be up for re-consideration? How does that kind of thinking relate to the mathematical problem posing that you might engage in with your students?

## REFERENCES

Brown, S. I., Cooney, T. J. & Jones, D. (1990). Mathematics teacher education. In R. Houston (Ed.), *Handbook of research on teacher education.* (pp. 639–656). NY, Macmillan.

Paul, R. W. (1990). *Critical thinking: What every person needs to survive in a rapidly changing world.* Rohnert Park, CA: Center for Critical Thinking and Moral Critique, Sonoma State University.

# 6 A Problem Posing Approach to Biology Education

John R. Jungck

In a recent article entitled "Problem-posing physics: A conceptual approach," Wytze Brouwer (1984) suggests the following approach to science education: (1) an "interactional style of teaching" should *pose* questions on the conceptional foundations that students have about a subject; (2) "student misconceptions can often be related to earlier, once perfectly respectable, historical preconceptions about" the physical and biological world; (3) "the teacher must have a profound respect for the people with whom he interacts, a faith in their own ability to analyze, to develop, or recreate their conceptions of reality to become more fully active as human beings"; (4) "apply the new explanatory models in familiar and new situations"; and (5) "test scientific understanding both conceptually and quantitatively."

In 1962, Thomas S. Kuhn's *The Structure of Scientific Revolutions* transformed many scholars' fundamental understanding of how science works by instantiating the role of actual historical events in the transformation of science rather than the previous role of history, which had been reconstructed for philosophical purposes. In this book, Kuhn gave credence to the current nature of scientific textbooks as playing the fundamental role in educating students to be prepared to do "normal" science and, in this sense, Kuhn could be viewed as having justified the normative nature of scientific textbooks. On the other hand, I believe that the current scientific textbooks are usually inadequate (although I still use them in most of my classes), needing compensation in four major ways. This is because textbooks usually describe science only in the philosophically reconstructed form that Kuhn was criticizing. Textbooks rarely devote much space to helping students understand (1) why earlier scientific conceptions are incompatible with our observations, (2) how to solve problems, (3) what heuristics (not the so-called scientific method) are generally useful in recognizing types of and approaches to problems, (4) who is capable of solving such problems, and, especially, (5) whose interests are served and what solutions are possible if a problem is posed in a particular fashion. Truman Schwartz (1977) goes even further:

> I am not convinced that the production of normal scientists should be our highest goal. I resonate with the words of Karl Popper when he says: "In my view the normal scientist as Kuhn describes him is a person one ought to be sorry for . . . The normal scientist in my view has been taught badly . . . He has been taught in a dogmatic spirit; he is a victim of indoctrination. He has learned a technique which can be applied without asking the reason why.". . . I hasten to add that I am not such an impractical romantic as to spurn all normal science. After all, chemistry has been phenomenally successful at solving puzzles which were safely within the boundary conditions imposed by accepted doctrine. Most of us seldom, if ever, venture beyond these limits. What I do submit is that an honest, unbowdlerized historical approach—one which admits of error, approximation, and human foibles—is not necessarily detrimental to conventional chemistry. Moreover, it stands a chance of generating the imagination and insight required for scientific revolutions.

But back to textbooks. I believe textbooks can be compensated in the following six ways: (1) pose questions to students on how they think things work, on why they think they work that way, and on the consequences of thinking the way they do; (2) encourage students to carry out independent investigations as much as possible; (3) develop computer software that builds experimental universes which provide students with professional tools to develop long-term research-like strategies for scientific problem solving in a very short time; (4) encourage students to work in groups, think aloud, evaluate each other's work, and try problem solving daily; (5) articulate clearly why and how recent research developments in cognitive psychology, science education, artificial intelligence, and history, philosophy and sociology of science, as well as some classical aspects of logic and epistemology, relate to learning how to solve problems; and finally, (6) students should actively engage in *posing* problems rather than simply solving those problems which are formulated in the back of chapters of textbooks. Only by recognizing the variations in and prejudices which we each bring to posing problems are students likely to be sensitive to their own culturally bound "theory laden observations" (Kuhn 1962).

How could we determine if many of the ideas that we have developed are culturally bound to Western science education? We need perspective on how students approach problem posing and problem solving in different cultures. Paolo Freire's *Pedagogy of the Oppressed* (1970) states his belief that "problem posing education" inherently will lead to a more humanizing and liberating knowledge:

> In problem-posing education, men develop their power to perceive critically the way they exist in the world with which and in which they find themselves; they come to see the world not as a static reality, but as a reality in process, in transformation. Although the dialectical relations of men with the world exist independently of how these relations are perceived (or whether or not they are perceived at all), it is true that the form of action men adopt is to a large extent a function of how they perceive themselves in the world. (pp. 70–71)

[*and*]

> Students, as they are increasingly posed with problems relating themselves in the world and with the world, will feel increasingly challenged and obliged to respond to that challenge. Because they apprehend the challenge as interrelated to other problems within a total context, not as a theoretical question, the resulting comprehension tends to be increasingly critical and thus constantly less alienated. Their response to the challenge evokes new challenges, followed by new understandings; and gradually the students come to regard themselves as committed.

Lecturing in an interactional style that attempts to draw out students, encouraging them to become active learners, is a minimal approach to a problem posing approach. Facilitation of learning rather than didactic pedagogy may be effected through encouraging cooperation via group learning, collaborative labs and projects to develop student interaction in scientific investigation, and by being highly problem-directed rather than descriptive.

Will this approach work well in a different cultural context because the approach depends on identifying problems of interest to the students, encouraging the students to develop their own approaches to problems, and getting students to work with one another? In addition, if we develop a large number of physical materials which are visual and tangible rather than linguistic, will these help develop scientific reasoning? How can we play the role of resource persons rather than disseminators of a fixed, rigid form of knowledge?

How can curriculum be developed and/or adapted which would be consonant with a problem posing approach? First, it would seem that problem posing curricula would be marked by both flexibility and specificity. Students should be allowed considerable freedom in choosing concentrations at both the high school and the undergraduate levels. Can concentrations be defined whereby students learn both breadth and a sequential series of skills? Brown and Walter (1983), two mathematics educators, note that concentrations are not defined by their content but instead by the problems they pose:

> The centrality of problem posing or question asking is picked up by Stephen Toulmin in his effort to understand how disciplines are subdivided within the sciences. What distinguishes atomic physics from molecular biology, for example? He points out that our first inclination to look for differences in the specific content is mistaken, for specific theories and concepts are transitory and certainly change over time. On the other hand, Toulmin comments:
>
>> If we mark sciences off from one another . . . by their respective "domains," even these domains have to be identified not by the types of objects with which they deal, but rather by the questions which arise about them . . . Any particular type of object will fall in the domain of (say) "biochemistry," only in so far as it is a topic for correspondingly "biochemical" questions.

An even deeper appreciation for the role of problem generation in literature is expressed by Mr. Lurie to his son, in Chaim Potok's novel *In the Beginning:*

> I want to tell you something my brother David, may he rest in peace, once said to me. He said it is as important to learn the important questions as it is the important answers. It is especially important to learn the questions to which there may not be good answers.

Indeed, we need to find out why some questions may not have good answers.

If we return to my critique of science textbooks, I believe that this is where they fail us most; namely, by focusing on facts and problems formulated in a "normal" science tradition (paradigm), we fail our students most by not raising the questions to which we have no good answers. I furthermore would argue that we fail to attract certain students to science because they see science as fixed with all of its major problems already solved rather than a dynamically growing approach to difficult problems, some of which have not yet been even articulated. Dorothy Buerk (1982) states this well:

> Many students of mathematics believe that the subject is only rules to be memorized, skills to be practiced, and methods to be followed precisely. I would like to propose that, for some adults, this view is not consistent with their more relativistic view of knowledge in general; this discrepancy in world view and view of mathematics causes discomfort with mathematics.

Can we write science books that are concerned with the unknown which lies within the range of "The Art of the Soluble?" By using prestated problems (dilemmas), we restrict our attention to some of the less exciting avenues of science for our students; furthermore, we "de-skill" our students. Daniel Pekarsky (1980) notes a similar situation in moral education:

> To know how to solve a moral dilemma once it has been laid out is one thing; to be able to identify the morally relevant features of an everyday situation and thus become aware that there is a moral problem, and to do so in a way that does justice to the complexity of the situation, is quite another. A program in moral education that takes predesignated moral dilemmas as its starting-point fails to take seriously enough the dispositions and skills that are necessary if the morally problematic is to be uncovered in the midst of the everyday.

Let me take the liberty of simply modifying Pekarsky's last sentence to show its exact equivalence in science education: A program in science education that takes predesignated scientific problems as its starting-point fails to take seriously enough the dispositions and skills that are necessary if the scientific problematic is to be uncovered in the midst of the everyday. This problem posing approach seems to be a *sine qua non* for developing students critical of "normal" science in the Kuhnian sense and capable of asking questions about matters of concern to their immediate lives.

I agree with Hilary Rose (1983) that a good curriculum builds the brain, the hands, and the heart. Thus, it is important to combine an intellectually rigorous cognitive development program with the craftsmanship of scientific tool use and affective growth in moral judgment (Brown 1984). I am very pragmatic in this domain because I believe that curriculum must emerge from local needs and personnel. Thus, national curricular policies have to be locally adapted and developed, but not compromised. Effective local implementation involves the active involvement and interaction of teachers, students and context. Field testing with feedback modifications is the most powerful and lasting way to transform policy into action at the local level.

Inasmuch as the problem posing approach referenced in this paper has dealt primarily with reading, mathematics, and physics education, why hasn't it been applied to biology which is concerned with life and death matters? The ecological and evolutionary complexity of our problems is enormous and has tremendous consequences. Let us solicit the help of our students in posing the problems of future scientific agendas. Problem posing and problem solving are inseparable activities (Walter and Brown 1977). Research doesn't indicate, experiments don't suggest, evidence doesn't show, and data does not imply. These anthropomorphisms hide their authors' intentions and prejudices. A problem posing approach which makes the inferences in a direct manner where the authors are explicit about their role in drawing these inferences seems a more honest approach to communicating science to students and involving students in science. What are the questions that we should be asking in biology education?

## REFERENCES

Anonymous. (1978). *Starting with Nina: The politics of learning.* (A film which "explores the socio-learning theories of Paulo Freire). D.E.C. Films.

Brouwer, W. (1984). Problem posing physics: A conceptual approach. *American Journal of Physics, 52*(7), 602–607.

Brown, S. I. (1984). The logic of problem generation: From morality and solving to de-posing and rebellion. *For the Learning of Mathematics, 4*(1), 9–20.

Brown, S. I. & Walter, M. I. (1983). *The art of problem posing.* Philadelphia: The Franklin Institute Press.

Buerk, D. (1982). An experience with some able women who avoid mathematics. *For the Learning of Mathematics, 3*(2), 19–24.

Freire, P. (1970). *Pedagogy of the oppressed.* New York: The Seabury Press.

Kuhn, T. S. (1962). *The structure of scientific revolutions.* Chicago: University of Chicago Press.

Pekarsky, D. (1980). Moral dilemmas and moral education. *Theory and Research in Social Education, 8*(1), 1–8.

Rose, H. (1983). Hand, brain, and heart: A feminist epistemology for the natural sciences. *Signs: Journal of Women in Culture and Society, 9*(11), 73–90.

Schwartz, A. T. (1977, August). The history of chemistry: Education for revolution. *Journal of Chemical Education, 54,* 467–468.

Walter, W. I. & Brown, S. I. (1977). Problem posing & problem solving: An illustration of their interdependence. *Mathematics Teacher, 70*(1), 4–13.

# 7 An Experience with Some Able Women Who Avoid Mathematics

Dorothy Buerk

And on the eighth day, God created mathematics. He took stainless steel, and he rolled it out thin, and he made it into a fence forty cubits high, and infinite cubits long. And on this fence, in fair capitals, he did print rules, theorems, axioms and pointed reminders. 'Invert and multiply.' 'The square on the hypotenuse is three decibels louder than one hand clapping.' 'Always do what's in the parentheses first.' And when he was finished, he said 'On one side of this fence will reside those who are good at math. And on the other will remain those who are bad at math, and woe unto them, for they shall weep and gnash their teeth.'

Math does make me think of a stainless steel wall—hard, cold, smooth, offering no handhold, all it does is glint back at me. Edge up to it, put your nose against it, it doesn't give anything back, you can't put a dent in it, it doesn't take your shape, it doesn't have any smell, all it does is make your nose cold. I like the shine of it—it does look smart, intelligent in an icy way. But I resent its cold impenetrability, its supercilious glare.

Many students of mathematics believe that the subject is only rules to be memorized, skills to be practiced, and methods to be followed precisely. I would like to propose that, for some adults, this view is not consistent with their more relativistic view of knowledge in general; this discrepancy in world view and view of mathematics causes discomfort with mathematics; and closing the gap between these disparate world views can make these adults feel more comfortable in approaching mathematics.

70

## LENSES TO VIEW MATHEMATICS

I have spoken to many people, especially to mature women who avoid mathematics or feel apprehensive about it. Many discuss experiences with timed tests or flash cards that led to embarrassment, an embarrassment that can be felt even years later when recalling the events. Many believed that if they were good at mathematics, then they should be very quick and competitive with it, and the failure to be quick and competitive led to feelings of inadequacy—especially if they were able people who thought reflectively. Many simply felt powerless in the face of mathematics because, as one woman put it, "the wicked mathematician has all the answers in the back of the book" [see Potter, 1978 for more of her statement].

Their experiences and feelings indicate a conception of mathematical knowledge that is termed "dualistic." I use "dualistic" as it is used by William G. Perry, Jr. [1970, 1981] in his developmental scheme describing how adults view knowledge (See Note 1). He defines dualism as:

> Division of meaning into two realms — Good versus Bad, Right versus Wrong, We versus They. All that is not Success is Failure, and the like. Right Answers exist *somewhere* for every problem, and authorities know them. Right Answers are to be memorized by hard work. Knowledge is quantitative. Agency is experienced as "out there" in Authority, test scores, the Right Job. [Perry, 1981, p. 79]

Several women with whom I have worked in a small group setting express this dualistic perception of mathematical knowledge vividly, and also the discomfort that comes with it. One expression of this view opens this paper and two others follow:

> I think of math problems or situations as having right and wrong answers (very black and white), but having a variety of ways to reach the answer. Unfortunately, my math teachers never stressed the fact that there could be more than one way to approach a problem. For this reason, and there are other reasons, I do not see math as a "creative activity." It is most definitely not linked to language, or music, or the other humanities. *Sonya* (See note 2)

> It is encouraging to me, when reading (about mathematics) to see the acceptance or existence of unanswered questions and "puzzlement." I feel that some of the "pressure" is taken off me to produce THE ANSWER via THE METHOD. *Sophie*

Must mathematics be viewed dualistically? Is mathematics only a collection of correct answers and proper methods? It is clear that this is not the view of those who write in *For the Learning of Mathematics.* The mathematics discussed in these pages evolves through a dynamic process that is exciting to those who discuss and experience it. For example, Henderson [1981] believes "that mathematics has meaning that can be experienced and imagined." Brown [1981] states:

> I believe that it is a serious error to conceptualize of (sic) mathematics as anything other than a human enterprise which among other things helps to clarify who we are and what we value.

In contrast to the experiences of these FLM writers, the women with whom I worked have experienced mathematics in a dualistic mode. They see it as a discipline that is rigid, removed, aloof, and without human ties, rather than one that is being discovered and developed. It is a collection of answers rather than a dynamic process that is alive and changing. The Authorities, the mathematicians, are mistrusted and suspect.

The women quoted above do not view other areas of their lives and experiences in this same dualistic way. General data from each of them have been rated at the positions Perry calls "relativism," in which one accepts all knowledge and all values as relative and contextual:

> Comparison, involving systems of logic, assumptions, and inferences, all relative to context, will show some interpretations to be "better," others "worse," many worthless. Yet even after extensive analysis there will remain areas of great concern in which reasonable people will reasonably disagree. It is in this sense that relativism is inescapable and forms the epistemological context of all further developments. [Perry, 1981, p. 88]

The acceptance of relativism as the way the world is, is a major—in fact, drastic—change. It is a change from stressing quantity to stressing quality, from relying on Authority, who has the answers and passes them on, to interacting with authority as an expert who can help in the intellectual search. It is a shift from an external source for action, for power, and for responsibility, to an internal one. A relativist becomes caught up in the excitement of ideas, the interrelationships of ideas, and an urge to play around with ideas. Asking questions and listening to the ideas of others is easier, but much learning is an active process in which the learner remains open to ideas, is self-processing, and even initiates the exploration. Care is taken to be precise in thought, to reason systematically, and to keep the context in focus. The security of relativism comes in this careful exploration of alternatives in many different areas of life.

The insecurity of relativism centers around the variety of alternatives and the realization that choices need to be made in some of these areas. Movement beyond relativism in Perry's scheme comes with the awareness that the individual herself/himself is the only one who can make these commitments. At issue now is responsibility, the individual's responsibility for personal commitments as a means of orienting himself/herself in a relativistic world. Perry calls this "Commitment in Relativism."

The discrepancy between the way these women viewed mathematical knowledge and the way they viewed knowledge in general both puzzled and disturbed me. How could women who were so able and intellectually mature as these women happened to be, seriously view mathematics in such a dualistic way? More importantly, could these women be helped to see mathematics from a perspective more closely aligned with their view of the world in general?

## THE STUDY

To answer the latter question I designed a study [Buerk 1981] in which five women, with general data rated as relativistic in Perry's scheme but retaining dualistic beliefs about the nature of mathematical knowledge, shared as a small group in mathematical experiences designed to help them to see the discipline of mathematics from a new perspective. I chose experiences and presented them in ways that would encourage growth through successive positions in the Perry scheme. A particular emphasis in the sessions was placed on "experiencing" a problem or question individually before discussing it as a group. When discussion did ensue its focus was on the question rather than an answer. The women were encouraged to ask questions about the meaning of the problem, to clarify any puzzling terms, and to share the mental images that the problem brought to mind. I believe that this "experiencing" step was important since it allowed each woman to make the problem meaningful for herself and to clarify it both visually and verbally. Once each woman "saw" the problem, resolution became the focus.

In addition, I asked the women to reflect on statements about mathematics and challenged them to articulate their own perceptions of mathematics, of mathematicians, and of the nature of the work mathematicians do. These reflections, and also their reactions and responses to the five sessions, were recorded in a journal which circulated among the participants during the 10 to 21 day intervals between the sessions. In addition, the women were interviewed before and after the sessions to determine their background and past experience with mathematics, and to allow them to express their feelings about mathematics.

## THE EXPERIENCE OF FIVE WOMEN

The first mathematical experience of the group was the "hand-shaking" question which I first thought about in a mathematics education course taught by Marion Walter at the State University of New York at Buffalo. I presented it to the five women in the following way:

> If the six of us wanted to meet by shaking each others' hands, how would you envision the number of handshakes?

Please, before you, the reader, read further, stop and think about this question. How would *you* envision the number of handshakes? Try to envision it in more than one way. Think about it for a moment. What questions follow for you as you think about "hand-shaking?"

I chose this experience because it is a question that can be approached in many ways, but which for most people does not bring to mind a formula or model from traditional classroom mathematics. The problem allows for a diversity of methods to reach THE ANSWER. It builds confidence since most people (with encouragement) do have a response. It also has some ambiguity which will allow some people to interpret it in ways that may yield more than more answer. (Is Maria's shaking of Sophie's hand the same as Sophie's shaking of Maria's hand? Is a right-handed handshake the same as a left-handed handshake? Can I shake my own hand?)

The hand-shaking question was posed to the women in the group, each of whom had a unique approach to it. Some approaches included:

*The process of elimination.*   Person 1 shakes 5 hands, person 2 has shaken hands with person 1 and has only 4 handshakes so: 5 + 4 + 3 + 2 + 1 + 0 = 15 handshakes. This approach of Maria was very insightful, but she wished she had come up with a visual one.

*The diagram approach.*   Here Sonya represented each person as a dot and each handshake as a line between two persons. A discussion arose about ways to count the lines.

*The grid approach.*    In this drawing Emmy represented the people along the outside and each box represented a handshake. Shaded are handshakes with one's self. The squares above the diagonal duplicate those below the diagonal. This was an extremely helpful visualization, but Emmy didn't trust it.

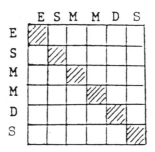

*The square dance approach.*    Mary's approach was to fantasize a square dance, choosing to count every handshake by considering person 1 shaking person 2's hand as different from person 2 shaking person 1's hand. Therefore, each of 6 people make 5 handshakes or 30 handshakes altogether. She then delighted the group with her ideas about styles of handshakes — firm, limp, joke, bonecrushing, and left-handed.

Once the methods were shared, it became apparent that each woman felt a strong sense of insecurity about her own approach. Comments made included the following:

I wasn't very mathematical about it.
It didn't occur to me to make a diagram.
I think that your way is 'better' than mine.

In general, group members did not exude confidence in their own methods.

Members raised many questions and gradually shifted the focus to some questions that might grow out of the initial problem. Suppose there were forty-five people. Which of our methods would work? Which would be convenient? Which would be efficient? How do you add the integers from 1 to 45 without getting bored? How could you make a formula to represent the problem? What form would a formula take? Is there more than one formula? Some comparison of methods, especially in terms of similarities and differences, ensued and served as a first approximation of "what mathematicians do."

The hand-shaking question involved these women in mathematics by first asking them to bring their own meaning to the question at hand. Each, on her own, could visualize the problem in her own way. The fact that Mary defined a handshake differently from the others led to a discussion of the need to agree on basic notions in a problem.

Following a second session which was on symbolizing, the third one, "tree-planting," was presented to the women in the following way:

> Imagine a flat plane—a geometric plane—completely flat. Does it have color? Weight? Can you see it? Plant 100 trees on your plane. Plant them in perfect rows and columns. You can have as many or as few as you like in any given row or column. Do you have a special kind of tree you are planting? These are very special trees of whatever variety you choose, because they are very tall and infinitely thin. They are perfectly straight. Are they all planted? Okay, please chop down one of the trees—any one at all. Replace it with yourself. As you stand on the stump you become infinitely thin as well. From wherever you put yourself, can you see out of your forest? Or, from your perspective and without bending to look around trees, can you see all of your trees? If not, how many can you see? You may look in as many directions as you like. [Adapted from a problem in Copes, 1980.]

Again, as reader of this paper, stop and fantasize your own forest. Once you have it clearly visualized place your trees on graph paper. From the place you placed yourself, can you see out? Suppose that you planted an infinite number of trees. Could you see out of your forest then? [See Copes, 1980]

This experience was chosen to give participants the chance to create their own problems. Each participant had at least three choices to make: the arrangement of the 100 trees, the positioning of the self, and the choice between two questions. If the person was on the outside edge of her forest, then the question of seeing out became trivial; the other question was "better" and more challenging. The answers to both questions were clearly dependent on the forest that each woman chose. The experience lends itself well to a "multiplistic" conception of mathematical knowledge. "Multiplicism" is the name given to the positions in Perry's Scheme that lie between "dualism" and "relativism." It is described by Perry [1981] as:

> Diversity of opinion and values is recognized as legitimate in areas where right answers are not yet known. Opinions remain atomistic without pattern or system. No judgments can be made among them so "everyone has a right to his own opinion; none can be called wrong."

In the first three sessions, the women were encouraged to adopt a multiplistic conception of mathematical knowledge. They were encouraged to see a problem from their own perspectives and to pursue their own directions. Individual ideas were encouraged and reinforced. Every attempt was made to build confidence, to talk in terms other than right and wrong, and to avoid traditional mathematical jargon. This direction was chosen to promote growth away from a dualistic conception of mathematical knowledge. It was also chosen to begin to compensate for the ways in which the dualistic perspective may have caused the feelings of apprehension about mathematics that these women have experienced: (1) by reinforcing an expectation of "needing to be quick," (2) by causing them to reject their intuition and with it their self-confidence in mathematics, and/or (3) by creating confusion about the role of conceptualization in an area viewed as dualistic.

In the fourth and fifth sessions, more stress was placed on supportive evidence, on evaluating, and on looking for "better" ways. These sessions were designed to encourage the relativistic view of mathematics.

The experience of the fourth gathering proved the most powerful of the five. Please experience it for yourself as you read it and resolve it to your own satisfaction before you read on. If it is new to you be prepared to be surprised. It was presented to the five women as follows:

> I'm going to present you with a question about the world. Please focus on your first "gutlevel" answer. Then let's talk together about the question to be sure that we really "see" it. The focus is to be on understanding the question. Try not to focus on a method to "solve it." Think again about what your intuition tells you. Keep a note of your intuitions. They need not be shared.

> The question is about the earth. Think of the equator. Put a flexible steel belt around the earth at the equator so that it follows exactly the contours of the earth. Now add 40 feet to the length of that belt and arrange it so that the belt is above the equator for its entire length. The belt still follows the contours of the earth at the equator and is raised above the equator by the same distance at every point. The model of a monorail track is a very nice one. (Thanks to Mary.)

> The question is, what will fit between the earth and the belt? That is, what is the distance between the earth and the belt? [Adapted from Stephen Brown, SUNY Buffalo Math Education course.]

I chose this fourth experience to generate a conflict, since for most people, this problem provides a situation in which the intuition and the theory (logic, data) are inconsistent. Once a conflict arises between the intuition and the data, what are the ways that people choose to deal with the inconsistency? I have seen it lead some to a relativistic view of mathematical knowledge and others to the crisis that leads them to take the responsibility for making commitments within the framework of a relativistic world.

I posed the "belted earth" problem with instructions to focus first on an intuitive answer rather than on a method of solution. Materials available and visible included: an old basketball, a rough-surfaced dyed orange, a ball of pie dough trying to flatten, an encyclopedia with the earth's measurements, and a calculator. I directed discussion through the following stages:

1. What is your initial intuition? Time was spent encouraging each person to become comfortable with her own perception about the size of the space between the earth and the belt. Most thought that it was minuscule, although Maria first thought that her hand could fit in the space and Mary saw some space because she envisioned the belt as a monorail track.
2. What questions do you have about the problem? Care was taken to encourage questions and clarify what the problem was really saying. The belt was viewed as a loose bracelet, a tight thing stretched, a monorail track. Members wondered if it got wet from ocean waves; if it followed the ocean bottom or the ocean surface, and if the belt conformed to the ocean at high tide or low tide. They wondered how it could be kept uniformly distant from the surface of a constantly moving earth. Did the 40 feet mean that the belt was 40 feet away from the earth? Could a circle be the model? What effect would the changes in the contours of the earth have? Members decided that the changes in the contours would average out and that a circle was a good model. A discussion of $\pi$ evolved. This clarification process was extremely important in order to get each person in touch with the problem in some depth and to be sure that the members were addressing the same problem.
3. What theory applies? The problem was envisioned with the following diagram where:
   $C$ = the circumference of the earth in feet.
   $C + 40$ = the circumference of the belt in feet.

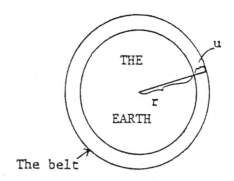

$r$ = the radius of the earth in feet.
$u$ (for unknown) = the increase in the radius in feet.

$\pi$ = the ratio of the circumference to the diameter in any circle.

$C = 2\pi r$ represents the circumference of the earth.

$C + 40 = 2\pi (r + u)$ represents the circumference of the belt.

$C + 40 = 2\pi r + 2\pi u$, but $C = 2\pi r$.

$C + 40 = C + 2u$ ($C$ can now be subtracted from both sides).

$40 = 2\pi u$

$u = 40/2\pi = 20/\pi \sim 20/3.14 \sim 6.37$ ft.

($\sim$ indicates an approximate value)

Therefore the belt stands a little more than 6 feet above the earth.

4. Participants were aware of a conflict. They were asked, "How do you resolve a conflict between your intuition and theory (data, logic)?" Some of the comments following the finding of $u \sim 6$ feet include:

Maria:  That's too much. I don't believe that.

Sonya:  I deal with the conflict by asking someone else.

Emmy:  The circumference of the earth *has* to matter.

Maria:  By using $C$, we've made it so the circumference could be 2 feet or 30,000 feet — as if it would all come out the same — and it can't!

Sonya:  Is $\pi$ in feet? Is 6.37 feet right? Why are we getting the wrong answer?

Sophie:  Let's try it for the orange. If the diameter is 3 inches that would be one-quarter of a foot.

Throughout this discussion, Mary was busy with the calculator, pencil, and paper. She assumed a radius of 2 feet and went through the computation. $u = 6.37$ feet! She assumed the orange had a diameter of 3 inches and got $u = 6.37$ feet! She felt triumphant in reporting that the only way to change $u = 6.37$ feet is to change the 40 feet.

Sophie pursued her quest for more data. She didn't want to use the numbers for the diameter and circumference of the world but wanted to use numbers. She talked about the orange. Finally, she went to the blackboard to pursue the question using the basketball. She assumed its circumference was 2 feet. She went through the computations on the board and found that $u$ (the distance the band stands above the basketball at every point on its "equator") was 6.37 feet.

The session ended with Mary feeling triumphant, not only because she had discovered a concept, but also because her intuition had been a monorail track above the earth. The others were all in conflict. Sonya was asking how $\pi$ could remain constant. Maria, too, was questioning the use of $\pi$ and not believing that 6 feet could be an answer. Sophie was excited by how well she could breeze through the computations on the board, but frustrated because she did not understand the answer. Emmy was confused. She wanted to believe her intuition, but could not find a way to reject the theory.

The session left members feeling more confident in their skills in using the algebra, and yet in conflict about its result.

That conflict, coming after they had gained some confidence and courage to try mathematical questions and to share their own ideas, was a powerful vehicle for changing their conception of mathematical knowledge.

To illustrate that power let's look at Maria's responses. Her fourth journal entry began with her experience of making the belted earth problem her own:

> Circumference problem. I first visualized the stainless steel band, shiny, about 5-6″ in width sloping across the earth, over grasses, down under marshes, under water. Wondered how long it would stay there before it began to look worn, like scrap metal. Seemed a perfect image for what mathematics does—take something organic and colorful and constrict it with something metallic and silver and rigidly out of place. (Why couldn't we have put a woven rush braid around the earth?)

It continued with a description of her conflict concluding with the following:

> So the problem must be in $C = 2 \pi r$—that you can't use $\pi$ as a fixed and actual "number." This is what it comes down to, but I'm not even convinced. The real problem lies with equating one $C + 40'$ with any *other* $C + 40'$ but I'm not sure how to solve it. *Still* believe *a* formula will apply, but can't see what I'm misapplying. *Still* trusting my first instinctive visualization. No exit.

In her final interview Maria indicated that she saw mathematics as a process and realized that speculation was necessary for the process to evolve. She concluded that:

> I feel like I have given birth to this new little creature, "math," and I have to take it home with me and where is it going to fit into my life now?

Note that Maria at this stage no longer sees mathematics as dualistic. She sees it as a process. She still wants to resolve her conflict in connection with the belted earth problem, but she also realizes that the responsibility is now hers and she must choose how, when, and if she will relate to mathematics. She states:

> Need to know if I "let math into my life," what would I *do* with it? (I have an image of opening my screen door, and a short furry 2-legged creature trotting in . . .)

Maria did not learn many new mathematical skills in five one-hour sessions, but she came to see mathematics from a different point of view. With that new perspective on mathematics she has taken responsibility for her own relationship to mathematics. Should she choose to take a course, she will not be without apprehension, but it will be an apprehension that she can deal with more readily than the apprehension that comes with the feelings of powerlessness because "the wicked mathematician has all the answers in the back of the book."

## SUMMARY

Many people who have moved beyond a dualistic view of knowledge in general, have difficulty with mathematics because they continue to see it only dualistically. My experience with the women in my study indicates that a more relativistic view of mathematics, coupled with a sense of personal responsibility for their own learning, made mathematics more approachable for them.

I am not making a general claim. Clearly my small and specialized sample is much too limited for that. I share this experience instead because of its depth and richness. There is much we can learn from listening to the way our students view mathematics as a field of knowledge. This group of able women provide us with a particularly articulate voice.

I share this experience also because I believe that a look at apprehension about mathematics from the perspective of a developmental scheme like William Perry's provides an alternative not usually present in work with the math avoidant. It is unfortunate that the term "math anxiety" has become such a popular one for it carries such a negative connotation.

Let me conclude with an excerpt from Emmy's final journal entry:

> In this group I have very much enjoyed the philosophic insights the (statements and journal) entries gave. They sent me off into regions I love, and I apologize (particularly to Maria) that those regions were sometimes strange and irritating to others. I guess I got carried away. I do feel however that this wandering has helped me to think through the values that define my own character as well as

mathematics. This seems to me an important step in repairing the flawed "relationship" I have with math. To think philosophically about math and to *play* with math problems simultaneously seem to me a lovely teaching method. For the first time I think I have had the experience of learning something about myself from doing math. I feel somehow relieved that we all do not have the same symptoms or sources of "math anxiety": the fact that my problems are particular makes them seem easier to overcome. I almost feel foolish enough to give calculus another whirl without a pre-calc. review course! I am suddenly very curious, though, about how to find out the math books and teachers that will teach to my strengths (visual intuition) and not to my weaknesses (mechanics: arithmetic, algebra, trig., etc.). If mathematical thinking can take on so many styles, so can math teaching. I have not seen so many math teaching styles, though, at least not consciously.

## NOTES

1. For a detailed description of Perry's scheme and its application to mathematics see the recent article by Larry Copes, "The Perry Development Scheme: a metaphor for learning and teaching mathematics", *For the Learning of Mathematics,* 3, 1: 38–44
2. The names that I have given to my subjects are the names of women prominent in the history of mathematics:

*Sonya* Kovalevskaya (1850–1891) — Russian
*Sophie* Germain (1776–1831) — French
*Emmy* Noether (1882–1935) — German
*Maria* Agnesi (1718–1799) — Italian
*Mary* Somerville (1780–1872) — English

## REFERENCES

Brown, S. I. (1981). Ye shall be known by your generations. *For the Learning of Mathematics, 1*(3), 27–36.

Buerk, D. (1981). *Changing the conception of mathematical knowledge in intellectually able, math avoidant women.* Doctoral dissertation, State University of New York at Buffalo.

Copes, L. (1980). *College teaching, mathematics, and the Perry development scheme.* Unpublished manuscript, MN: Institute for Studies in Educational Mathematics.

Henderson, D. (1981). Three papers. *For the Learning of Mathematics, 1*(3), 12–15.

Perry, W. G., Jr. (1970). *Forms of intellectual and ethical development in the college years: A scheme.* NY: Holt, Rinehart and Winston.

Perry, W. G., Jr. (1981). Cognitive and ethical growth: the making of meaning. In A. Chickering (Ed.), *The modern American college.* San Francisco: Jossey Bass, 76–116.

Potter, B. D. (1978). The train and the fly or why I hate math. In Letters to the Editor, *The Two-Year College Journal, 9*(1), 3–4.

# 8
# The Invisible Hand Operating in Mathematics Instruction: Students' Conceptions and Expectations

Raffaella Borasi

Although competence in mathematics is becoming a prerequisite for most careers, an increasing number of students seem unable to succeed in mathematics. *Mathematics anxiety* is widespread, and too many students avoid enrolling in mathematics courses unless they are strictly required. Although social and psychological factors are certainly at the root of this situation, the role played by students' views of school mathematics could also be crucial.

Consider, for example, the following image of mathematics, expressed by a bright math-avoidant woman (Buerk 1981) and supported by many of the high school students to whom it was shown (Borasi 1986b):

> Math does make me think of a stainless steel wall — hard, cold, smooth, offering no handhold, all it does is glint back at me. Edge up to it, put your nose against it, it doesn't give anything back, you can't put a dent in it, it doesn't take your shape, it doesn't have any smell, all it does is make your nose cold. I like the shine of it — it does look smart, intelligent in an icy way. But I resent its cold impenetrability, its supercilious glare.

It is not surprising that persons holding such a perception of mathematics — as a rigid and impersonal discipline — will avoid engaging in mathematical activities, regardless of their ability! Students' conceptions and expectations, however, can influence their everyday approach to mathematics learning in less obvious yet powerful ways, as the following episodes illustrate.

One of my greatest surprises as a teacher and a researcher occurred at the

83

beginning of an experimental eleventh-grade mathematics course, which I designed and taught in a local alternative high school. I was proud of the plan for my first class, which was to be devoted to the discussion of *why* the outrageous simplification $\frac{16}{64} = \frac{1}{4}$ gives a correct result—an intriguing problem that requires genuine problem solving and invites reflection about the nature of equations and methods for their solution. I also felt that the class had gone reasonably well, since the students had been attentive, and a few of them had shown considerable creativity in their approach to the problem. Little did I know what lay in store for me! I learned the next day that many of my students, far from being intrigued by my approach, were actually planning to drop the course! The best I could do (once a first reaction of panic was subdued) was to have a frank talk with the students in our next class. There their reaction became suddenly clear as they expressed their deep concerns through a deceptively simple question: "How can we ever cover the course material if we spend a *whole* class on just *one* problem?" This question revealed the asymmetry in expectations that I as a teacher and they as a class shared about what makes a mathematical activity valuable. Such a discrepancy certainly went a long way in aborting the success of my lesson, regardless of the pedagogical merit of its design.

The experience described above is in no way unique to the introduction of experimental teaching techniques, as shown by the following episode, which occurred in a regular class and dealt with a more *traditional* curriculum content:

> Our class had been discussing the coordinate system, three points along with their coordinates were drawn on the board, and the students were asked to determine if these points were collinear. After the definition of collinearity had been recalled and discussed, there was a long silence broken finally by a student who said that he did not know the formula for collinear. When the explanation was made that no such formula existed, he replied, almost in anger, "How do you expect us to do a problem if we've never been shown how?" The instructor replied, "Think about it." All gave up except one student. He responded (a bit hesitantly) that he did not know how to do the problem *mathematically* but he did think he could do it his own way, and proceeded to reason that if *A, B, C* were three points (in that order) on a straight line, then the distance from *A* to *B* added to the distance from *B* to *C* must be the same as the distance from *A* to *C;* otherwise they were not collinear points and formed a triangle. Of course, the teacher was delighted with his answer and asked why he did not think it was mathematical. His reply was: "Because I thought of it myself." (Oaks 1987, pp. 3–4)

In this episode, the students' implicit assumption that mathematics consists of a predetermined set of rules and procedures "passed on" by teachers to the next generation did not allow them to consider *thinking on their own* as an appropriate strategy to approach mathematical problems.

As illustrated by these examples, students' conceptions of the nature of mathematics and their expectations with respect to school mathematics can constitute a powerful force operating "behind the scenes" in any mathematics class. Since these beliefs are usually deep-seated and unconscious, it is unfortunately difficult for teachers to access them, and even more so to attempt to modify those conceptions that may seem counterproductive, as revealed by several research studies on people's belief systems (see, for example, Perry 1970; Cooney 1985; Brown and Cooney 1987). Just "telling" the students what mathematics really is and what is expected from them is not likely to do much to resolve the problem.

With the goal of helping teachers deal constructively with their students' views of school mathematics, I have attempted in this article to identify some beliefs that could prove dysfunctional to students' learning of mathematics and to discuss ways to help students gain a better appreciation of the nature of mathematics and consequently develop expectations and behaviors that are more conducive to success.

## SOME "DYSFUNCTIONAL" MATHEMATICAL BELIEFS

The results of several recent research studies have shed light on students' views about school mathematics (see, for example, Buerk 1981, 1985; Oaks 1987; Schoenfeld 1985). To "capture" even unconsciously held mathematical beliefs, a variety of methodologies were employed, including a combination of open-ended questionnaires, in-depth interviews, student journals, videotapes of problem-solving sessions, and even the analysis of metaphors of mathematics created by students. These studies have revealed the *existence* of a set of beliefs that could negatively affect mathematics instruction. These beliefs are briefly synthesized below—with some inevitable oversimplications—with respect to four key categories:

- *The scope of mathematical activity:* Providing the correct answer to given problems, which are always well defined and have exact and predetermined solutions. This applies to the activity of both mathematicians and mathematics students, though the complexity of the problems approached would obviously differ.
- *The nature of mathematical activity:* Appropriately recalling and applying learned procedures to solve given problems.
- *The nature of mathematical knowledge:* In mathematics, everything is either right or wrong; there are no gray areas where personal judgment, taste, or values can play a role. This applies both to the facts and procedures that constitute the body of mathematics and to the results of each individual's mathematical activity.

• *The origin of mathematical knowledge:* Mathematics always existed as a finished product; at best, mathematicians at times discover and reveal some new parts of it, while each generation of students "absorb" the finished products as they are transmitted to them.

In what follows, for brevity, let us refer to the view of mathematics characterized by this set of beliefs as "dualistic." I would like now to support my original claim that the mathematical beliefs identified above may prove dysfunctional for their holders' learning of mathematics.

First of all, these beliefs reflect a limited appreciation of the nature of mathematics. They are oblivious to the struggle and creativity that was required to achieve even what may now be considered the most basic mathematical results. The presence of controversy on some mathematical issues, and in particular on the foundations of the discipline, is not even conceived as possible. Ignored is the role played by value judgments in the creation of new definitions, the choice of an axiomatic system, or the evaluation of alternative proofs for the same theorem. These beliefs also fail to capture the complexity and nonlinearity of mathematical applications.

Even accepting that a dualistic view of mathematics does not reflect the nature of the discipline, one might still wonder whether it really represents a problem for mathematics students, especially at an early stage of schooling. After all, a sense of security may even be derived from perceiving a discipline as perfectly organized and void of ambiguity. And there could be the hope that a student's limited conception of mathematics will modify and expand naturally as more advanced mathematical topics are encountered.

Unfortunately, this may be true at best only for some of the most able students. For the majority of the students, a dualistic view of mathematics is more likely to cause expectations and behaviors leading to anxiety and academic failure. For example, the students' reactions in the episodes reported earlier can now be easily explained, and could even have been predicted, as a direct consequence of these mathematical beliefs. And so are other ineffective student behaviors that usually puzzle mathematics teachers, such as focusing on memorization rather than on conceptual understanding or lacking constructive strategies to cope with learning difficulties. Additional examples of counterproductive expectations and behaviors that could be derived as "corollaries" from a dualistic view of mathematics can be found in figure 8.1.

What can have caused students to develop such mathematical beliefs, since these beliefs do not truly reflect the nature of mathematics and certainly cannot be justified on the basis of facilitating the learning of mathematics? Social stereotypes and stages of intellectual development may

## Some Common Misconceptions and Their Explanation in Terms of a Dualistic View of Mathematics

**Learning mathematics is a straightforward matter and practice *alone* should "make perfect.".**

If mathematical activity is equated with applying the appropriate algorithms to given problems, then learning mathematics should only involve taking down notes of the procedures that the teacher gives out, memorizing all the steps in their correct sequence, and practicing on a sufficient number of exercises so as to become able to perform the procedures quickly and without mistakes; and if the results are not satisfactory, all you can do is practice more.

**It is not good trying to reason things out on your own.**

If you learned successfully, you should be able to do the problem quickly; if instead you did not pay sufficient attention to the teacher in class or did not practice enough, "thinking" alone cannot help remedy these deficiencies; furthermore, since for any problem there is a correct procedure to be applied and you do not know it, you cannot hope to come up with another one on your own.

**Staying too much on a problem is a waste of time.**

It is difficult to appreciate that by looking closely at *one* problem, you can learn something more general and transferable, when "reasoning it out" is not conceived as appropriate and furthermore there is the perception that mathematical results are disjoint.

**You cannot learn from your mistakes.**

If there is not connection between a right and a wrong way of doing mathematics, trying to analyze and understand your mistakes is just a waste of time; therefore, if you do something wrong, you should forget about it and start back from scratch to do it right.

**Formal mathematics is just a frill.**

Proofs, deductions, formal definitions, ar not really helpful when it comes to finding and correctly applying the appropriate algorithm to solve a problem; therefore, formal mathematics becomes a ritual that can be performed on the teacher's request but that can be ignored in the context of solving problems.

**History and philosophy of mathematics are irrelevant to learning mathematics.**

A reasonable conclusion, if mathematics learning is defined in terms of mastering procedures to solve specific problems and such procedures are believed to exist independently of the way they were discovered; consequently, readings and teachers' lectures in these areas are likely to be perceived as a digression and a "luxury," which a student struggling with mathematics cannot afford.

**A good teacher should never confuse you.**

From the assumption that ambiguity does not exist in mathematics, it follows that the origin of confusion must be a poor presentation; this expectation can easily justify "blaming the teacher" for the learning difficulties experienced and also make students resistant to innovative teaching approaches based on discovery, problem solving, and explorations.

FIG. 8.1.

certainly play a role in shaping students' conceptions, as suggested by research on world views such as those of Perry (1970) and Belenky et al. (1986). Yet another cause could be the way mathematics is presented in school. We should not be surprised at mathematics students' overwhelming concern with product and answers when the most important measure of academic success is given by the score received on standardized multiple-choice tests taken under considerable time pressures. Nor can we ignore the fact that a simplistic view of mathematics learning could simply reflect what goes on, almost every day, in most mathematics classes: the teacher introduces a new concept or rule by lecture, applies it in a few simplistic examples, assigns similar exercises for practice to the students for home-work, later on goes over these homework results, and finally verifies the students' ability to perform the same task in a test situation.

## STRATEGIES TO HELP STUDENTS RECONCEIVE THEIR VIEWS OF MATHEMATICS

The previous analysis of some dysfunctional conceptions of mathematics has made clear that helping students reconceive their views of the discipline should emerge as a major concern for mathematics teaching. At the same time, the challenge of such a task has also been revealed. Because of the very nature of a dualistic and product-focused view of mathematics, we can in fact expect that a direct approach through lectures and readings *about* mathematics will do little good, since students are likely to perceive such initiatives as merely "icing on the cake" and consequently pay little attention to them. Nevertheless, the task is not impossible, and valuable insights and strategies to approach it have been suggested both by research on how conceptions develop (such as Perry 1970; Oaks 1987; Brown and Cooney 1987) and by interventions that actively attempted to affect people's conceptions (such as Buerk 1981; Borasi 1986a, 1988).

All these studies have pointed out the key importance for students to *become aware of, and reflect on, their beliefs, as well as possible alternatives,* since beliefs are more powerful the more they are held unconscious and unquestioned. They have also made us aware of the danger of rhetoric: in abstract discussions on the nature of mathematics and its learning there is too often the risk of cross-talk and misunderstanding because students and teachers may not attribute the same meaning even to key terms such as *understanding* or *mathematical problem.* It is important, therefore, to offer students opportunities to *engage in mathematical activities that can gen-erate doubt* in assumptions taken for granted up to that point and to *personally experience more humanistic aspects of mathematics* of which they may have been unaware.

For example, some cognitive conflict could be stimulated by presenting students with unsolvable mathematical paradoxes and contradictions, for which even "Authority" does not know the answer; inherent limitations, as in attempting to define $0^0$; statements that may be right or wrong depending on the context, such as "multiplication is repeated addition." All these experiences, in fact, are in conflict with the dualistic expectation that mathematics is a perfect domain, where everything is either right or wrong, and may consequently lead students to reconsider the validity of such beliefs. In turn, encouraging students to share their procedures to perform a specific mathematical task could shake the belief that the correct result in mathematics can be reached in only one way and pave the way for the acceptance of alternatives in other aspects of mathematics as well. A better appreciation for the role of personal judgment and values in mathematical activity could instead be fostered by engaging students in the study of problem situations (National Council of Teachers of Mathematics 1989), where the solver is required to define the problem more precisely, select relevant information, and evaluate the acceptability of the solution(s) reached. Problem-posing activities, where students are encouraged to generate on their own mathematical questions worthy of study, could also greatly contribute to this goal (Brown and Walter 1983).

An awareness of critical events in the history of mathematics, and of the troubled genesis of some specific topics and results, could well complement the previous experiences by showing how even great mathematicians had to struggle and use considerable ingenuity to produce what is now accepted within the body of mathematical knowledge.

Although the strategies described above can help teachers design activities that may generate healthy doubt in their students' minds, it is also important that the time and opportunity to *reflect on these experiences* be provided. Class discussions are certainly a good means to do so, but there are considerable advantages to complementing them with expressive writing activities. Writing assignments requiring students to report and reflect on their mathematical experiences, perhaps with the help of a few thought-provoking questions by the teacher, will in fact force each individual student to take a stance, can provide more time and leisure to identify, work out, and express satisfactorily one's ideas, and finally will produce a written product that could be exchanged with others and provide a record of development over time (see, for example, Borasi and Rose [in press] and the article by Azzolino in this volume).

It is also important, however, that the activities thus described do not remain isolated episodes within an essentially traditional mathematics curriculum. Rather, throughout the curriculum the students should experience aspects of mathematics and mathematical activity that are consistent with a nondualistic view of the discipline. This requires that potential problems, controversial points, and the possibility of alternative interpretations be continuously highlighted and discussed rather than hidden from students' consideration, even at the risk of occasionally losing some clarity. Students should also be put in a position in which they can engage in the creation of original answers, the generation of new questions, and the evaluation of their own mathematical activity instead of always being immediately furnished with the final and "correct" product, even if this process will require more time and consequently the quantity of mathematical results covered in the curriculum may need to be reduced.

Obviously, the strategies described in this section cannot be seen as simple additions to the existing curriculum and traditional teaching approaches; rather, they require that we radically reconsider the goals, content, and means of mathematics instruction for the 1990s.

## CONCLUSIONS

In this article I have examined the nature and consequences of students' holding a dualistic view of mathematics—that is, the belief that mathematics is a collection of disjoint, predetermined, and absolutely correct facts and procedures that are used to solve specific problems and that teachers are supposed to pass on such facts and procedures to students, who in turn will memorize them for later recall and application for the solution of given problems. It has been argued that students holding such a view are likely to be passive learners, focusing on memorization rather than conceptual understanding, ignorant of the need of making personal meaning of the material presented in class, lacking constructive strategies to deal with learning difficulties, and therefore liable to experience limited success in school mathematics.

If this is true in the context of the current curriculum, which still mostly stresses the acquisition of technical facts and skills, it will be true even more so in the future. The new goals for school mathematics in the 1990s, put forth by the National Council of Teachers of Mathematics to better respond to the needs of our ever-changing and increasingly technological world, are in fact even more at odds with a dualistic view of mathematics, since they stress the importance for students of "(1) becoming a mathematical problem solver; (2) learning to communicate mathematically; (3) learning to reason mathematically; (4) valuing mathematics; (5) becoming confident in one's ability to do mathematics" (National Council of Teachers of Mathematics 1989, p. 12).

Overcoming the problem of students' inappropriate conceptions of mathematics should thus become a priority for mathematics instruction in the 1990s. Yet, this article has shown that this is no easy task. We cannot hope to resolve the problem by adding an isolated unit on "The Nature of Mathematics." Rather, teachers will need to create a variety of learning situations throughout the curriculum, which will lead students to become aware of, and question, their perceptions of mathematics as they experience and reflect on humanistic aspects of mathematics and its learning. It is hoped that this article has provided the stimulus and encouragement, as well as some concrete suggestions, to engage in such a challenging and worthwhile enterprise.

## REFERENCES

Belenky, M. F., Clinchy, B. M., Goldberger, N. R., & Tarule, J. M. (1986). *Women's Ways of Knowing: The development of self, voice, and mind.* New York: Basic Books.

Borasi, R. (1986a). *On the educational uses of mathematical errors.* Unpublished doctoral dissertation, State University of New York at Buffalo.

Borasi, R. (1986b). Behind the scenes. *Mathematics Teaching. 117,* 38–39.

Borasi, R. (1988). *Using errors as springboards to inquire into the nature of mathematical definitions: A teaching experiment.* (Report No. MDR8651528). National Science Foundation.

Borasi, R., & Rose, B. (1989) Journal Writing and Mathematics Instruction. *Educational Studies in Mathematics, 20,* 347–365.

Brown, S. I., & Cooney, T. J. (1988). Stalking the dualism between theory and practice. In P. F. L. Verstappen (Ed.), *Proceedings of second conference on systematic cooperation between theory and practice in mathematics education* (pp. 21–40). The Netherlands: National Institute for Curriculum Development.

Brown, S. I. & Walter, M. I. (1990). *The art of problem posing* (2 Ed.). Hillsdale, NJ: Lawrence Erlbaum Associates.

Buerk, D. (1981). *Changing the conception of mathematical knowledge of intellectually able math avoidant women.* Unpublished doctoral dissertation, State University of New York at Buffalo.

Buerk, D. (1985). The voices of women making meaning in mathematics. *Journal of Education, 167*(3), 59–70.

Cooney, T. J. (1985). A beginning teacher's view of problem solving. *Journal for Research in Mathematics Education, 16,* 324–326.

National Council of Teachers of Mathematics. (1989). *Curriculum and evaluation standards for school mathematics.* Reston, VA: Author.

Oaks, A. (1987). *The effect of the interaction of mathematics and affective constructs on college students in remedial mathematics,* Unpublished doctoral dissertation, University of Rochester.

Perry, W. G. (1970). *Intellectual and ethical development in the college years: A scheme.* New York: Holt, Rinehart & Winston.

Schoenfeld, A. (1985). *Mathematical problem solving.* Orlando, FL: Academic Press.

# The Logic of Problem Generation: From Morality and Solving to De-Posing and Rebellion

## 9

Stephen I. Brown

## TWO CULTURES

A quarter of a century ago, C. P. Snow accurately pointed out how little the two cultures—roughly the sciences and the humanities—have learned to understand each other and to gain from the wisdom they each have to offer. [Snow, 1959]

> Between the two a gulf of mutual incomprehension—sometimes . . . hostility and dislike, but most of all lack of understanding [emerges]. They have a curious distorted image of each other . . . non-scientists tend to think of scientists as brash and boastful. . . . [They] have a rooted impression that the scientists are shallowly optimistic, unaware of man's condition. On the other hand, the scientists believe that the literary intellectuals are totally lacking in foresight, peculiarly unconcerned with their brother men, in a deep sense anti-intellectual, anxious to restrict both art and thought to the existential moment. [p. 12]

Not only are their problem solving styles different, but more importantly there are divergent views on what it means for something to be solved. It is worth observing that as a profession, mathematics education is almost by definition bound to the schizophrenic state of searching for and creating the "snow-capped" bridges; for mathematics is more closely aligned with the culture and world view of science and education with the humanities.

As we search for a better understanding of what problem solving might be about, however, we have not only neglected to build bridges, but we have tended to ignore most non-mathematical educational terrain that might be worth connecting in the first place.

In particular, we have overlooked those educational efforts in other fields which have been concerned with problem solving but have indicated that concern through a different language. Dewey's analysis of "reflective thought" and of the concept of "intelligence" would seem to offer a rich complement to much of the problem solving rhetoric. The role of *doubt, surprise* and *habit* in problem solving explored by Dewey would seem to complement much of the influential work of Polya, and would offer options we have not yet incorporated in much of our thinking about problem solving in the curriculum. [Dewey, 1920, 1933]

We have much to learn about the role of dialogue in problem solving, something we in mathematics education have tended to view in pale "discovery exercise" terms at best. Yet the use and analysis of dialogue in educational settings has been the hallmark not only of English education, but of several curriculum programs in other fields as well. "Public controversy" in the social studies in the late 60's and early 70's was a central theme around which students were taught not only to carry on intelligent dialogue, but more importantly to unearth and to discuss controversial and sometimes incompatible points of view. [Oliver and Newman, 1970] It would enrich considerably what it is we call problem solving in mathematics, if we were to entertain the possibility that for logical as well as pedagogical reasons, we might encourage not merely complementary, but incompatible perspectives on a problem or a series of problems. Furthermore such curriculum in the social studies as well as in the newly emerging field of philosophy for children might enable us to help students *appreciate* irreconcilable differences rather than to resolve or dissolve them as we are prone to do in mathematics. [Lipman, et al., 1977]

"Critical" thinking is another "near relative" of problem solving that began influencing the curriculum in schools as far back as the progressive education era, and there is a considerable history of efforts to integrate different disciplines through the use of critical thought. [Taba, 1950] It is a history that is worth understanding not only because of its connection with problem solving, but because the theme is presently undergoing rejuvenation in the non-scientific disciplines much as problem solving has re-emerged in mathematics and science.

## MORAL EDUCATION AND KOHLBERG

In closing this section, we turn towards one area within which the tunes of critical thinking have been re-sung recently—that of moral education. The issues that emerge here and those that we develop in the next section are part of the new (and not yet well integrated) backdrop mentioned in the first section.

First of all, we might ask why critical thinking and moral education have been joined at all. To many people, they would seem to occupy different poles. The connection hinges on our concern for the teaching of values in a pluralistic, democratic society. How do we go about such education in a public school setting without indoctrinating with regard to a particular religious or ethnic point of view? Though we might argue over whether or not critical thinking is a set of values itself (and if so, why such a collection is more neutral than other religious points of view), the liberal tradition of thinking critically about whatever values one adopts does provide an entrée for those concerned with morality in a pluralistic society.

Though there a number of different kinds of programs within which moral education is taught [Lickona, 1976], most of them rely heavily upon contrived or natural dilemmas as a starting point. Our focus here will be on Kohlberg's program of moral development and education. A typical dilemma he has used for much of his research and for his deliberate program of education as well is the Heinz dilemma:

> In Europe, a woman was near death from a rare form of cancer. There was one drug that the doctors thought might save her, a form of radium that a druggist in the same town had recently discovered. The druggist was charging $2000, ten times what the drug cost him to make. The sick woman's husband, Heinz, went to everyone he knew to borrow the money, but he could only get together about half of what the drug cost. He told the druggist that his wife was dying and asked him to sell it cheaper or let him pay later. But the druggist said, "no." So Heinz got desperate and broke into the man's store to steal the drug for his wife. [Kohlberg, 1976, p. 42]

Should Heinz have stolen the drug? Based upon an analysis of longitudinal case studies to answers of dilemmas of this sort, Kohlberg has created a scheme of moral growth that he claims is developmental. Furthermore, he has created not only a research tool but an educational program around such dilemmas. It is through discussing and justifying responses to such dilemmas that students mature in their ability to find good reasons for their choices.

It is not the specific value that one chooses (e.g., steal the drug vs. allow the wife to die), but the reasons offered for the decision that places people along a scale of moral development.

At the lowest level of moral maturity, (pre-conventional) Kohlberg finds that people argue primarily from an awareness of punishment and reward. Thus someone at a lowest stage of development might claim that Heinz should not steal the drug because he would be punished by being sent to jail, or he might claim that he should steal it because his wife might pay him well for doing so. It is almost as if the punishment inheres in the action itself. At a later stage (conventional), people argue from the more abstract perspective of what is expected of you and also from the point of view of the need to maintain law and order. At the highest stage of principled morality, one argues on the basis not of rules that could conceivably change but with regard for abstract principles of justice and respect for the dignity of human beings. Such principles single out fairness and impartiality as part of the very definition of morality.

None of these structural arguments (e.g., punishment/reward, law and order, justice) in themselves dictate what is a correct resolution of any dilemma. Rather they form part of the web that is used to justify the decisions made, and it is in listening to these reasons that Kohlberg and his followers are capable of deciding upon one's level of moral development.

## GILLIGAN'S CHALLENGE

Despite the fact that Kohlberg's scheme for negotiating moral development neglects to focus upon action, it is a refreshing counterpoint to a program of moral education which conceives of its role as one of inculcating specific values in the absence of reason. Nevertheless, there has been some penetrating criticism of his scheme recently — a criticism which condemns much of Kohlberg's work on grounds of sexism. That is, Kohlberg's research and ultimately his scheme for what represents a correct hierarchy of development is based upon his longitudinal research *only with males*. Once the scheme was created and the stages developmentally construed, Kohlberg interviewed females and concluded that their deviation from the established hierarchical scheme implied an arrested form of moral development.

Gilligan [1982] points out that the existence of a totally different category scheme for men and women not only may be a consequence of different psychological dynamics, but rather than exhibiting a logically inferior mind set, it suggests moral categories that are desperately in need of incorporation with those already derived. Compare the following two responses to the Heinz dilemma, one by Jake, an eleven-year old boy and the second by Amy, an eleven-year old girl. Jake is clear that Heinz should steal the drug at the outset, and justifies his choice as follows:

> For one thing a human life is worth more than money, and if the druggist makes only $1000, he is still going to live, but if Heinz doesn't steal the drug, his wife is going to die. (Why is life worth more than money?) Because the druggist can get a thousand dollars later from rich people with cancer, but

Heinz can't get his wife again. (Why not?) Because people are all different and
so you couldn't get Heinz's wife again. [Gilligan, 1982, p. 26]

Amy on the other hand equivocates in responding to whether or not Heinz
should steal the drug:

> Well, I don't think so. I think there might be other ways besides stealing it,
> like if he could borrow the money or make a loan or something, but he really
> shouldn't steal the drug — but his wife shouldn't die either. If he stole the drug,
> he might save his wife then, but if he did, he might have to go to jail, and then
> his wife might get sicker again, and he couldn't get more of the drug, and it
> might not be good. So, they should really just talk it out and find some other
> way to make the money. [p. 28]

Notice that Jake *accepts* the dilemma and begins to argue over the
relationship of property to life. Amy, on the other hand, is less interested in
property and focuses more on the interpersonal dynamics among the
characters. More importantly, Amy refuses to accept the dilemma as it is
stated, but is searching for some less polarized and less of a zero sum game.
Kohlberg's interpretation of such a response would imply that Amy does
not have a mature understanding of the nature of the moral issue
involved — that she neglects to appreciate that this hypothetical case is
attempting to test the sense in which the subject appreciates that in a moral
scheme life takes precedence over property. Gilligan on the other hand in
analyzing a large number of such responses has concluded not that the
females are arrested in their ability to move through his developmental
scheme, but that they tend to abide by a system which is orthogonal to that
developed by Kohlberg — a system within which the concepts of *caring* and
*responsibility* rather than *justice* and *rights* ripen over time.
Gilligan [1982] comments with regard to Amy's response:

> Her world is a world of relationships and psychological truths where an
> awareness of the connection between people gives rise to a recognition of
> responsibility for one another, a perception of the need for response. Seen in
> this light, her understanding of morality as arising from the recognition of
> relationship, her belief in communication as the mode of conflict resolution,
> and her conviction that the solution of the dilemma will follow from its
> compelling representation seem far from naive or cognitively immature. [p.
> 30]

The difference between a "Kohlbergian" and a "Gilliganish" conception of
morality is well captured by two different adult responses to the question,
"what does morality mean to you?" [Lyons, 1983] A man interviewed
comments:

Morality is basically having a reason for doing what's right, what one ought to do; and, when you are put in a situation where you have to choose from amongst alternatives, being able to recognize when there is an issue of "ought" at stake and when there is not; and then . . . having some reason for choosing among alternatives. [p. 125]

A woman interviewed on the same question comments:

Morality is a type of consciousness, I guess a sensitivity to humanity, that you can affect someone else's life. You can affect you own life and you have the responsibility not to endanger other people's lives or to hurt other people. So morality is complex. Morality is realizing that there is a play between self and others and that you are going to have to take responsibility for both of them. It's sort of a consciousness of your influence over what's going on. [p. 125]

While Gilligan and her associates do not claim that development is sex bound in such a way that the two systems are tightly partitioned according to gender, they do claim to have located a scheme that tends to be associated more readily with a female than a male voice. Behind the female voice of responsibility and caring, some of the following characteristics appear to me to surface:

1. A context-boundedness,
2. A disinclination to set general principles to be used in future cases,
3. A concern with connectedness among people.

Though not all of these characteristics are exhibited in Amy's response, they do appear in interviews with mature women. *Context-boundedness* represents a plea for more information that takes the form not only of requesting more details (e.g., what is the relationship between husband and wife?) but of searching for a way of locating the episode within a broader context. Thus unlike men, mature women might tend to respond not by trying to resolve the dilemma, but by exhibiting a sense of *indignation* that such a situation as the Heinz dilemma might arise in the first place. Such a response might take the following form: "The question you should be asking me is 'What are the horrendous circumstances that caused our society to evolve in such a way that dilemmas of this sort could even arise — that people have learned to miscommunicate so poorly'?"

The second characteristic I have isolated above, is an effort to attempt to understand each situation in a fresh light, rather than in a legalistic way — i.e., in terms of already established precedent. Connected with context boundedness it is the desire to see the fullness of "this" situation in order to see how it might be *different from* (and thus require new insight) rather than compatible with one that has already been settled.

With regard to the third characteristic, conflict is less a logical puzzle to be resolved but rather an indication of an unfortunate fracture in human relationships—something to be "mended" rather than an invitation for some judgement.

In the next section we turn towards a consideration, in a rather global way, of how it is that a Gilliganish perspective of morality might impinge on the study of mathematics. While we have not yet drawn any explicit links, it is not difficult to intuit not only that it threatens the status quo but that it sets a possible foundation for the relationship of problem generation to problem solving. Though we shall focus upon the findings from the field of moral education, we do not wish to lose sight of some of the other humanistic areas of curriculum from which mathematics education might derive enlightenment.

## Kohlberg versus Gilligan: The Transition From Solving To Posing

It surely appears that problem solving in mathematics education has been dominated by a Kohlbergian rather than a Gilliganish one. Gilligan herself has an intuition for such a proposition, when she comments with regard to Jake's response to the Heinz dilemma:

> Fascinated by the power of logic, this eleven-year old boy locates truth in math, which he says is "the only thing that is totally logical." Considering the moral dilemma to be "sort of like a math problem with humans," he sets it up as an equation and proceeds to work out the solution. Since his solution is rationally derived, he assumes that anyone following reason would arrive at the same conclusion and thus that a judge would consider stealing to be the right thing for Heinz to do. [p. 26–27]

The set of problems to be solved as well as the axioms and definitions to be woven into proofs are part of "the given"—the taken-for-granted reality upon which students are to operate. It is not only that the curriculum is "de-peopled" in that contexts and concepts are for the most part presented ahistorically and unproblematically, but as it is presently constituted the curriculum offers little encouragement for students to move beyond merely accepting the non-purposeful tasks.

Furthermore, rather than being encouraged to try to capture what may be *unique* and unrelated to previous established precedent in a given mathematical activity (the legalistic mode of thought we referred to as the second characteristic behind Gilligan's analysis of morality as responsibility and caring), much of the curriculum is presented as an "unfolding" so that one is "supposed" to see similarity rather than difference with past experience. It is commonplace surely in word problems to tell people to *ignore* rather than to embellish matters of detail on the ground that one is after the underlying structure and not the "noise" that inheres in the problem.

In so focusing on essential isomorphic features of structures, the curriculum tends not only to threaten a Gilliganish perspective, but as importantly, it supports only one half of what I perceive much of mathematics to be about. That is, mathematics not only is a search for what is essentially common among ostensibly different structures, but is as much an effort to reveal essential differences among structures that appear to be similar. [See Brown, 1982a]

With regard to context boundedness, there is essentially no curriculum that would encourage students to explicitly ask questions like:

What purpose is served by my solving this problem or this set of problems?

Why am I being asked to engage in this activity at this time?

What am I finding out about myself and others as a result of participating in this task?

How is the relationship of mathematics to society and culture illuminated by my studying how I or other people in the history of the discipline have viewed this phenomenon?

Elsewhere [Brown 1973, 1982] I have discussed how I first began to incorporate such reflection as part of my own mathematics teaching, and presently I shall have other illustrations. There are a number of serious questions that must be thought through, however, before one feels comfortable in encouraging the generation and reflection of the kinds of questions indicated above. We need to be asking ourselves whether or not that kind of reflection represents respectable mathematical thinking. In addition we ought to be concerned about the ability of students to handle that thinking in their early stages of mathematical development.

It is interesting to observe that though we are cajoled by many to integrate mathematics with other fields, the "real world" applications seem to be narrowly defined in terms of the scientific rather than the humanistic disciplines. In particular questions of value or ethics are essentially nonexistent. That is particularly surprising in light of the fact that a major rationale for relating mathematics to other fields seems to be that such activity may enable students to better solve "real world" problems that they encounter on their own. I know of essentially no "real world" problems that one decides to engage in for which there is not embedded some value implications.

McGinty and Meyerson [1980] suggest some steps one might want to take to develop curriculum for which value judgments are an explicit component. Consider a problem like the following:

> Suppose a bag of grass seed covers 400 square feet. How many bags would be needed to uniformly cover 1850 square feet? [p. 501]

So far so dull. It is not only that for many students the above would not constitute a problem, but more importantly it lacks any reasonable conception of context-boundedness. The authors, however, go on to suggest inquiry that is more "real worldish" that most of the word problems students encounter. They ask:

> Should the person buy 5 bags and save the leftover — figuring prices will rise next year? Buy 5 bags and spread it thicker? Buy 4 bags and spread it thinner? [p. 502]

Once we become aware of ethical/value questions as a central component of decision making, it is clear that there is much more we might do in the way of generating problems for students as well as encouraging them to do so on their own. One of the *au courant* curriculum areas is probability and statistics. As a profession, we correctly appreciate that we need to do more to prepare students to operate in an uncertain world, wherein one's fate is not sown with the kind of exactitude that much of the earlier curriculum has implied. In creating such a curriculum, however, we continue to give the false illusion that mathematical competence is all that is required to decide wisely. Compare *any* probability problem (selected at random of course) from any curriculum in mathematics with the following probability problem:

> A close relative of yours has been hit by an automobile. He has been unconscious for one month. The doctors have told you that unless he is operated upon, he will live but remain a vegetable for the rest of his life. They can perform an operation which, if successful, would restore his consciousness. They have determined, however, that the probability of being successful is .05, and if they fail in their effort to restore consciousness, he will certainly die.

What counsel would you give the doctors? One could clearly embed the above problem in a more challenging mathematical setting, for example, setting up the conditions that would have enabled one to arrive at the .05 probability (or perhaps modifying it so that outer limits are set on the probability of survival) but nevertheless, it is such ethical questions in many different forms that plague most thinking people as they go through life making decisions.

Is such a problem generation on the part of the teacher or student an ingredient of mathematical thought? I do not think the answer is clear. There is nothing god-given and written in stone that establishes what is and is not part of the domain of mathematics, and clearly what has constituted legitimate thinking in the discipline has changed considerably over time. Even if questions of the kind we have been raising in this section, however, would move us in directions that are at odds with the dominant and respectable mode of mathematical thought, it is worth appreciating that as educators we have a responsibility to future citizens that transcends our passing along *only* mathematical thought. The latter appears to me to be a very narrow view of what it means to educate. In realizing that only a very small percentage of our students will be mathematicians, we have not adequately explored our obligation to those who will not expand the field *per se*. We have mistakenly identified our task for the majority as one of "softening" an otherwise rigorous curriculum. What may be called for is an ever more intellectually demanding curriculum, but one in which mathematics is embedded in a web of concerns that are more "real world" oriented than any of us have begun to imagine.

Is it worth observing that such complication of mathematical thinking may in fact pose a major threat to a concept that we have begun in recent years to revere—that of mathematization. In attempting to find reason to believe that children can indeed function as mathematicians (as opposed to exhibiting routine imitative skills), David Wheeler [1982] looks towards exceptional cases of mathematical precocity. He comments:

> I don't see children however exceptional functioning as historians, or a lawyers, or as psychologists, for instance, since these are extremely complex functionings that involve subtle relationships between (sic) several frames of reference. But I would hypothesize that mathematics belongs with art, music, writing and possibly science, as one of a class of activities that require only a particular kind of response to be made by an individual to his immediate, direct experience. [p. 45]

While I would certainly not wish to pit mathematization, as Wheeler describes it, against the mindless symbol-pushing that represents its polar opposite, I believe that as educators we are obliged to push the bounds of complicating that discipline in an effort to engage the minds of students in directions that define their humanity.

## Down From a Crescendo

How do we descend from the heights and perhaps the overinflated language which concluded the previous section? Perhaps one way is to take stock of where we have been led and to try to sharpen the implication that might follow. The confrontation between Kohlberg and Gilligan has served two purposes that appear on the surface to be very different. First of all, we have used the challenge of Gilligan's research to point out that there is a world view that has achieved empirical expression with regard to issues of morality but which is worth taking seriously in other domains as well. Moving beneath the concepts of caring and responsibility established by Gilligan, we find dimensions that are not strictly moral in character but which deal with *purpose, situation specificity* (a non-legalistic mode) and *people connectedness.* We have suggested that very little of the existing mathematics curriculum caters to those characteristics, and in fact the dominant mode caters to their opposite.

Secondly, we have not only used Gilligan in contrast to Kohlberg to establish broad categories within which the present curriculum is deficient, but we have pointed out that what the two perspectives have in common — namely a concern with morality — represents a field of inquiry that may be as important to integrate with mathematical thinking as are the more standard disciplines that form the backbone of more conventional applications.

Both of these perspectives have potentially revolutionary implications. They not only suggest the need for both teacher and student to incorporate a more serious problem generating perspective (including the broad types of questions raised at the beginning of the previous section) as an essential ingredient of problem solving, but they have the potential to infect every aspect of mathematics education from drill and practice, to an understanding of underlying mathematical structures.

## REFERENCES

Brown, S. I. (1971). Rationality, irrationality and surprise. *Mathematics Teaching, 55,* 13–19.

Brown, S. I. (1973). Mathematics and humanistic themes: Sum considerations. *Educational Theory, 23,* 191–214.

Brown, S. I. (1974). Musing on multiplication. *Mathematics Teaching, 61,* 26–30.

Brown, S. I. (1975). A new multiplication algorithm: On the complexity of simplicity. *Arithmetic Teacher, 22,* 546–554.

Brown, S. I. (1976). From the golden rectangle and Fibonacci to pedagogy and problem posing. *Mathematics Teacher, 69,* 180–186.

Brown, S. I. (1978). *Some "prime" comparisons.* Reston, VA: National Council of Teachers of Mathematics.

Brown, S. I. (1979). Some limitations of the structure movement in mathematics education: The meaning of *why. Mathematics Gazette of Ontario, 17,* 35–40.

Brown, S. I. (1981). Ye shall be known by your generations. *For the Learning of Mathematics, 3,* 27–36.

Brown, S. I. (1982). On humanistic alternatives in the practice of teacher education. *Journal of Research and Development in Education, 15*(4), 1–12.

Brown, S. I. (1982a). Distributing isomorphic imagery. *New York State Mathematics Teachers Journal, 32,* 21–30.

Brown, S. I. (1982b). Problem posing: The problem generation gap, *Math Lab Matrix, 16,* 1–5.

Brown, S. I. & Walter, M. I. (1983). *The art of problem posing.* Philadelphia: The Franklin Institute Press.

Dewey, J. (1920). *Reconstruction in philosophy.* NY: Holt & Co.

Dewey, J. (1933). *How we think.* NY: Heath.

Getzels, J. W., & Jackson, P. W. (1961). *Creativity and intelligence: Explorations with gifted students.* NY: John Wiley & Sons.

Gilligan, C. (1982): *In a different voice.* Cambridge, MA: Harvard University Press.

Higginson, W. C. (1973). *Towards mathesis: A paradigm for the development of humanistic mathematics curricula.* Unpublished doctoral dissertation, University of Alberta.

Hofstadter, D. (1982). Meta-Mathemagical Themes: Variations on a theme as the essence of imagination. *Scientific American, 247*(4), 20–29.

Kohlberg, L. (1976). Moral stages and motivation: The cognitive developmental approach. In T. Lickona (Ed.). *Moral development and behavior: Theory, research and social issues* (pp. 31–53). NY: Holt, Rinehart & Winston.

Lickona, T. (Ed.). (1976). *Moral development and behavior: Theory, research and social issues.* NY: Holt, Rinehart & Winston.

Lipman, M., Sharp, A. S., & Oscanyan F. (1977). *Philosophy in the classroom.* Montclair, NJ: Institute for the Advancement of Philosophy for Children.

Lyons, N. P. (1983). Two perspective on self, relationships and morality. *Harvard Educational Review, 53,* 125–145.

McGinty, R. L. & Meyerson, L. N. (1980). Problem solving: Look beyond the right answer. *Mathematics Teacher, 73,* 501–503.

Oliver, D., & Newman, F. (1970). *Clarifying public controversy: An approach to teaching social studies.* Boston: Little, Brown.

Snow, C. P. (1959). *The two cultures: And a second look.* NY: The New American Library.

Taba, H. (1950). The problems in developing critical thinking. *Progressive Education, 26,* 45–48.

Walter, M. I., & Brown, S. I. (1969). What if not? *Mathematics Teaching, 46,* 38–45.

Walter, M. I., & Brown, S. I. (1977). Problem posing and problem solving: An illustration of their interdependence. *Mathematics Teacher, 70,* 4–13.

Wheeler, D. (1982). Mathematization matters. *For the Learning of Mathematics, 3*(1), 45–47.

# 10 Vice into Virtue, or Seven Deadly Sins of Education Redeemed

Israel Scheffler

My purpose in what follows is to reveal some of the virtues hidden in what are typically deemed unqualified educational vices. I am encouraged in this purpose by one of William James's (1958) celebrated *Talks to Teachers,* in which he urged his hearers not to disparage passions "often . . . considered unworthy . . . to appeal to in the young," but rather to redirect them to good educational use, "reaping [their] advantages . . . in such a way as to [achieve] a maximum of benefit with a minimum of harm." Thus, as against Rousseau, who in his *Emile* had attacked the use of rivalry as a motive in education, James defended "the feeling of rivalry" as lying "at the very basis of our being, all social improvement being largely due to it. There is a noble and generous kind of rivalry," James wrote, "as well as a spiteful and greedy kind; and the noble and generous form is particularly common in child-hood. All games owe the zest which they bring with them to the fact that they are rooted in the emulous passion, yet they are the chief means of training in fairness and magnanimity. Can the teacher afford to throw such an ally away?" (pp. 49–51).

Similarly, James defended appeal to the pupil's pride and pugnacity since, "in their more refined and noble forms they play a great part in the schoolroom and in education generally, being in some characters most potent spurs to effort. Pugnacity," he continued,

> need not be thought of merely in the form of physical combativeness. It can be taken in the sense of a general unwillingness to be beaten by any kind of difficulty. It is what makes us feel "stumped" and challenged by arduous achievements, and is essential to a spirited and enterprising character. . . . It

is nonsense to suppose that every step in education *can* be interesting. The fighting impulse must often be appealed to. Make the pupil feel ashamed of being scared at fractions, of being "downed" by the law of falling bodies; rouse his pugnacity and pride, and he will rush at the difficult places with a sort of inner wrath at himself that is one of his best moral faculties. . . . The teacher who never rouses this sort of pugnacious excitement in his pupils falls short of one of his best forms of usefulness (pp. 51–52).

Finally, James had warm words for the sense of ownership, which he considered "also one of the radical endowments of the race." This sense, he wrote, "begins in the second year of life. Among the first words which an infant learns to utter are the words 'my' and 'mine' and woe to the parents of twins who fail to provide their gifts in duplicate." Private proprietorship James considered to be part of human nature, it being

essential to mental health that the individual should have something beyond the bare clothes on his back to which he can assert exclusive possession, and which he may defend adversely against the world. Even those religious orders who make the most stringent vows of poverty have found it necessary to relax the rule a little in favor of the human heart made unhappy by reduction to too disinterested terms. The monk must have his books; the nun must have her little garden, and the images and pictures in her room (p. 52).

In education, said James, ownership "can be appealed to in many ways," notably "in connection with one of its special forms of activity, the collecting impulse." Much of scholarship, indeed, rests on "bibliography, memory and erudition" and in these aspects owes its interest to "the collecting instinct" rather than "to our cravings after rationality."

A man wishes a complete collection of information, wishes to know more about a subject than anybody else, much as another may wish to own more dollars or more early editions . . . than anybody else.

The teacher who can work this impulse into the school tasks is fortunate. Almost all children collect something. A tactful teacher may get them to take pleasure in collecting books; in keeping a neat and orderly collection of notes; in starting, when they are mature enough, a card catalogue; in preserving every drawing or map which they make. Neatness, order, and method are thus instinctively gained, along with the other benefits which the possession of the collection entails (pp. 52–53).

In showing how rivalry, pugnacity, and ownership can be put to good use, James provokes us to reevaluate these motives. They are not to be denied or disparaged as such, but rather prized as potential instruments of teaching. Viewed as James invites us to view them, they are transformed from debits into assets, from obstacles into opportunities, from vices into potential virtues. The naive aversion they typically inspire is now seen to have been misguided; overcoming such aversion yields an enhancement of educational effectiveness.

Now James's idea of extracting the good from the presumptively bad can be applied well beyond the sphere of motives, to include educational states or processes generally. In what follows, I have selected seven of these for discussion. They are in such bad repute as to be popularly deemed vices — it is perhaps no exaggeration to say that they are regarded as the seven deadly sins of education. I will argue, to the contrary, that they are neither sinful nor deadly in themselves — nor are they vicious. Rather, when properly qualified, they constitute educational goods, to be wisely promoted or exploited by the sensitive teacher. These seven are ignorance, negativity, forgetting, guesswork, irrelevance, procrastination, and idleness.

I began first with ignorance.

## IGNORANCE, OR WHAT WE DO NOT KNOW

There is a quatrain that used to be circulated concerning the great nineteenth-century classicist and educator Benjamin Jowett, Master of Balliol College, Oxford, that went like this:

> My name is Benjamin Jowett
> If there is any knowledge I know it
> I am the Master of Balliol College
> If I don't know it, it's not knowledge.

The ideal represented in this quatrain is not only an ancient one but one that still reigns in education. It is knowledge that is thought to be the be-all and end-all of education, the gaining of fact and the eradication of ignorance. Knowledge, after all, is what education has to convey, the justification of schooling and the *raison d'être* of the teacher. Far from tempering this hoary conception, the momentum of our computer age with its associated notions of information, data bases, and the knowledge explosion serves only to entrench it still further in the public mind.

One flaw in this idea is its suggestion that the matter of education is available in advance, and that schooling is simply the process of conveying it to pupils. John Stuart Mill (1962) in a scathing comment of 1832 on the education of his day, characterized such education as "all *cram*. . . . The world already knows everything, and has only to tell it to its children" (p. 545). In its popular modern dress, the idea is that all the important information is already in the data base, the pupil needing only computer literacy to enable him or her to retrieve it at will (see Scheffler, 1986a for criticism of such conceptions of information).

A related defect is the confusion of *knowing* with the acquisition of *knowledge,* that is, reliable information. To know a proposition expressing a bit of information is, however, more than just to have accepted it. It is to have earned the right to accept it, through a grasp of its meaning and warrant. Knowledge as a collective heritage of recorded information is indeed a fundamental resource of the teacher but it cannot be transferred bit by bit in growing accumulation within the student's mind. The teacher must strive rather to promote an insight into the meaning, basis, and use of this collective heritage, so that the student may in fact come to know it rather than simply being informed of it. Even such knowing is not enough to express the aims of education, however, for it leaves out the opportunity for *innovation* by the learner, his ability to go beyond a knowing of available truths. We do not feed into the learner's mind all that we hope he will have as an end result of our teaching. We do *not* already know everything; our pupils can and should be expected to gain new understandings beyond our present grasp; they will need to revise our science, expand and modify our scholarship, recast our social, historical, and legal suppositions. They will, in short, need to discover new truths beyond our ken and reshape the very heritage of knowledge we now teach them (see Scheffler, 1973, p. 71 and Scheffler, 1965, Chapter 3).

It is here that the importance of acknowledging our ignorance can be clearly seen. To educate our pupils for innovation in the world of the future they will inhabit, we evidently cannot *give* them the new truths they will need, for such truths in the nature of the case are unavailable to us; they lie beyond our grasp. If we had them to give, they would not be new. What we *can* do is to avoid giving our pupils the idea that our heritage is a seamless web of settled facts, an idea that is not only false but intimidates the adventurous spirit and chills the impulse of inquiry.

The plain fact is that this heritage of ours has gaps and fissures, jagged edges and incomplete contours. Our science contains contradictions and enigmas, our philosophy deep difficulties and disagreements. It is not just that there is much we do not yet know. It is that we positively know our knowledge to be faulty. Our ignorance is not simply privative—indicating a lack—but throbbingly assertive. No one has expressed better the precarious nature of our heritage than Michael Oakeshott (1967), who describes it in these words:

> This inheritance is an historic achievement, . . . it is contingent upon circumstances, it is miscellaneous and incoherent; it is what human beings have achieved, not by the impulsion of a final cause, but by exploiting the

opportunities of fortune and by means of their own efforts. . . . It does not deliver to us a clear and unambiguous message; it speaks often in riddles; it offers us advice and suggestion, recommendations, aids to reflection, rather than directives. It has been put together, not by designers but by men who knew only dimly what they did (p. 162).

Acquiring such an inheritance is described by Oakshott (1966) as "learning how to participate in a conversation: it is at once initiation into an inheritance in which we have a life interest, and the exploration of its intimations" (p. 344).

The further point I would emphasize is that the gaps, difficulties, and riddles in our heritage are not educational obstacles to be minimized or deplored. They are in fact inspirational to the learner. They tell him that there are things he is called on to do — intimations to explore, enigmas to resolve, conflicts to overcome, revisions to effect, new paths to discover. Ignorance, thus interpreted, is not a mere void but an infinite space rich with educational possibilities to beckon the active young mind. Far from being an embarrassment, ignorance should be given a proud and central place in our curricula.

Socrates claimed to be the wisest of men only because, although he knew nothing, he also knew that he knew nothing. I do not take matters to that extreme. There are plenty of things we do know but what we know is precarious, gappy, enormously limited, and problematic. It is these aspects of our knowledge that should be made explicit and salient in our teaching.

## NEGATIVITY, OR THE POWER OF NEGATIVE THINKING

Positive thinking is much overrated. There is, for example, a prevalent myth that science builds its theories on positive results — but a theory is false if it fails anywhere, no matter how many positive results it yields. Superstitions are false not because they lack positive instances but because they also have negative ones; they are believed because we fasten on the former and conveniently repress the latter. As Karl Popper (1959) has emphasized, we ought not praise a theory simply because it has positive instances. A theory needs, further, to withstand our most strenuous efforts to overthrow it. It is the mark of science not that it seeks to confirm its theories but rather that it seeks to disconfirm them through experimental test, hunting for whatever negative instances they may harbor.

Negative thinking shows its power not only in the testing of ideas but also in their generation. New ideas thrive in the imagination, which negates what is and ponders what might be. Without the capacity to eliminate the positive and accentuate the negative, to skirt the actual and explore the possible, we would be forever captives of the past. Nothing is more important in education than finding ways to cultivate the imagination, the power to negate actualities in thought and leave pious pedantries behind.

Two former colleagues of mine, Stephen Brown and Marion Walter, several years ago worked out a way of teaching mathematics that has, ever since, exemplified for me the possibilities of teaching in this spirit. Given a mathematical system defined by various structural features, these colleagues ask their students to think away each of these features one by one and to explore the consequences of doing so, in imagination. The name Brown and Walter (Walter & Brown, 1969; Brown & Walter, 1983) give their scheme serves as an apt emblem of the powers of negative thinking: they call it "What if Not?"

## FORGETTING, OR LEST WE REMEMBER

Forgetting is commonly considered an educational defect, memory a virtue. We rarely stop to think what a blessing forgetting is and how disabling memory can be. The late psychologist A. R. Luria (1987) has given us, in his book *The Mind of a Mnemonist,* the case study of a man with enormous powers of memory, whose problem was the need to forget. "If," to take one example from Luria's account, "a passage were read to him quickly, one image would collide with another in his mind; images would begin to crowd in upon one another and would become contorted. How then was he to understand anything in this chaos of images?" (p. 112). Remembering too much tended to extinguish the abstract significance of the passage.

Now this is of course a quite extreme case, but our ordinary experience yields a variety of examples where the elimination of traces rather than their preservation is what is required. Take first the learning of skills, for example, skating, dancing, driving, swimming, typing, where the object is economy and grace of performance. Here suppression of our early halting steps is clearly functional; the traces of our stumblings, thrashings, and miscues are to be wiped out, the chisel marks smoothed away. If, as you type, you try to remember the clumsy movements of your fingers when you first learned, your present performance will be disrupted.

Even the initial rules for correct performance we may have been given are superseded and in time turn opaque. Max Black (1967) has described this process as one of "phenomenological compression" or "condensation" of the original formula. The experienced chess player, in his example, no longer uses the cumbersome rule he initially learned for the movement of the Knight. He "comes to *see* the target square as available for the Knight . . . and the criteria embodied in the original formula may be so effectively suppressed that 'verbal articulation' may be disconcertingly difficult." Such "intuitive transformation" of rules Black deems "of fundamental importance to educational method. . . . Whatever the topic—a mathematical proof, the conjugation of a verb, the salient features of the Industrial Revolution—the data must be 'rendered down,' simplified, structured, if they are to be assimilated, remembered and properly used." (pp. 100–101). Remembering these data under their new structure, it must be added, requires a progressive suppression of their initial structure. Remembering here builds on forgetting.

In areas of personal life, we often recognize the virtues of forgetting. To a friend who tends to preserve a memory of every slight or to cry over spilt milk, we may say "Forget it," thus acknowledging that memory is not always desirable. Yet in the realm of cognitive learning, we tend to overlook the parallel, that is, that some things deserve forgetting.

In these days of the so-called knowledge explosion, we have, it is true, been sensitized to the expansion of scholarly publication and the inevitable inability to take it all in. Luckily, to dignify each item published in the scholarly journals as "knowledge" is surely an exaggeration. Much, it is safe to say, is trivial, much is derivative, a good deal is worthless. Of what remains, only the tiniest fraction may bear on issues that concern a given learner, scholar, or professional. When we add to scholarly publication the torrent of words that bombards us from all sides in popular format, the situation becomes desperate.

Without the ability to ignore and forget, to turn a deaf ear to claims on our attention, we could develop no sustained cognitive efforts, no steady intellectual habits; constant distraction would be our lot. The urgent problem this situation presents to us is how to develop criteria for filtering the significant from the trivial, how to decide what to read and what to pass by, what to learn and what to ignore, what to remember and what to forget. This is at once a practical, an intellectual, and a moral problem that ought to be faced and dealt with in education. It is a problem that cannot readily be raised unless we recognize that to be forgotten is the proper fate of much that is communicated to us. To recognize the problem would, in any case, go far in mitigating the emphasis on examinations based on memory, which still reigns over so much of education's dominion.

Here again, William James (1958) is a sure guide when he writes:

> We are all too apt to measure the gains of our pupils by their proficiency in directly reproducing in a recitation or an examination such matters as they may have learned, and inarticulate power in them is something of which we always underestimate the value. . . . But this is a great mistake. It is but a small part of our experience in life that we are ever able articulately to recall. And yet the whole of it has had its influence in shaping our character and defining our tendencies to judge and act. Although the ready memory is a great blessing to its possessor, the vaguer memory of a subject, of having once had to do with it, of its neighborhood, and of where we may go to recover it

again, constitutes in most men and women the chief fruit of their education. This is true even in professional education. The doctor, the lawyer, are seldom able to decide upon a case off-hand. They differ from other men only through the fact that they know how to get at the materials for decision in five minutes or half an hour: whereas the layman is unable to get at the materials at all. . . .

Be patient, then, and sympathetic with the type of mind that cuts a poor figure in examinations. It may, in the long examination which life sets us, come out in the end in better shape than the glib and ready reproducer, its passions being deeper, its purposes more worthy, its combining power less commonplace, and its total mental output consequently more important (pp. 100–101).

## GUESSWORK, OR I GUESS SO

Guessing has long been deemed an academic defect. The pupil is supposed to know the right answer and not to guess at it. "Either you know it or you don't" is the attitude of innumerable teachers, parents, and test-makers, past and present, but it is no more realistic for being so widely shared. The fact is that the capacity for guessing is one of the most important — perhaps *the* most important — mental capacity we have, without which we would be so handicapped as not to be able to sustain our very lives.

In one of its senses guessing may be identified with estimation of some numerical value; in another and broader sense, it may be equated with theorizing. Both senses have in common that the guess goes beyond what is known or can be ascertained for sure. Take first estimation. You want to know the number of cookies in the jar but have no time to count them; you want to know the length of the bookshelf but cannot actually measure it; you want to decide how much money to take along on your trip in order to cover expenses. In no case do you have an answer that is certainly right, but that does not mean that every answer is equally wrong. There are better and worse ways to deal with such problems of estimation even if in the nature of the case they work by indirection and yield approximations ruled by probability. Probability is after all, as Bishop Butler (1900) said, the very guide of life. . . .

Now consider explanatory or predictive theories rather than numerical estimates. Our theories as to why things happen and what may be expected to happen set the context for all of life's activities. Not one step do we take, not one decision do we make that does not depend on such theories, yet their status is commonly misconceived. They are thought, in the case of science at least, to rest on a solid basis of fact. Some textbooks even speak of "theory construction" as if theories were systematically built on a firm foundation of factual evidence.

There simply is no systematic method for constructing theories. They are not in fact built but guessed. Science can indeed be described, in Michael Polanyi's (1964) words, as "a consistent effort at guessing" and "the propositions of science" as "in the nature of guesses" (pp. 23, 31). Evidence, to be sure, plays a critical role in the testing of scientific theories, but no theory can be tested that is not first formulated. That initial formulation goes well beyond all accumulated evidence. It is neither self-evident nor logically derived from the facts we have. Guesswork is its true source; its only method what Einstein (1949) described as "a free play with concepts" (p. 13). Not the eradication but the cultivation of guessing is the proper goal of education—guessing that is not simply the flipping of a coin but is responsive to the problem at hand and committed to the verdicts of ongoing tests. To build such guesswork into the educational experience of our students is a primary challenge to educators.

## IRRELEVANCE, OR WHAT IS THE POINT

That everything educational has a point is a popular dogma. In America, this is often interpreted as saying that every subject of the curriculum has to justify itself by its usefulness; each unit of teaching needs to be shown relevant to the future life of the student or of the nation if it is to be allowed a place in schooling. Everything in education is thus thrown on the defensive; it is guilty until proven innocent.

The effect is that some subjects are scanted or demoted in significance; others are simply distorted. Thus the arts are conceived as mere frill; history, in Henry Ford's immortal phrase, "is bunk" (Sward, 1948; p. 110); ethics is a luxury or an intrusion, while mathematics and the sciences are taught with primary emphasis on their technological applications. All of education becomes a handmaiden of utility, which rules the system.

Why is utility given such preeminent status? Is it obvious that everything has a point? What is the point of Beethoven's Rasoumovsky Quartets, of what use is the *Mona Lisa,* what is the usefulness of Fermat's Last Theorem, Aristotle's physics, Kant's *Fundamentals of the Metaphysic of Morals,* Plato's *Republic,* of Chaucer's *Canterbury Tales*? If having a point is so interpreted as to declare these works pointless, the dogma of relevance loses all credibility. On the other hand, if they are conceded to *have point* (that is to say significance, independently of having *a* point, that is, some useful application), the idea of excluding or demoting them in schooling is robbed of its force.

Such works do not gain their significance as means to the achievement of certain ends independently defined. Rather, they help to define the ends that confer significance on other things as means. Holding exemplars of value before the growing mind is justification enough for various elements of education, for the job of education is not only to provide persons with useful techniques but also to provide techniques with persons who have been made sensitive to the endless quest for knowledge and ideal values. If the fruit of schooling is its use in life, it must indeed be a life itself infused with a respect for knowledge and value (see Scheffler, 1973; p. 135).

## PROCRASTINATION, OR DO NOT DO IT NOW

Can anything good be said about procrastination, that favorite whipping boy of parents, educators, and efficiency experts? Yes, a good deal, on reflection. In any sensible ordering of one's affairs, some things need putting off. I refer not merely to the psychologists' concept of delayed gratification, the capacity for which is thought to be a sign of maturity, but also, first of all, to delayed worry, the capacity for which is a boon to mental health. When multiple cares crowd in on a person, often in the dead of a sleepless night but frequently as well in the heat of the day, the ability to sort these cares by temporal urgency is of enormous benefit. What I have in mind here is the ability to distribute one's several worries along a time line running into the indefinite future, thus relieving their combined pressure on the present and allowing them to be dealt with one by one. Scarlett O'Hara's "I'll think about that tomorrow" is often a good policy.

Second, there is the putting off of various tasks and duties in the interest primarily of efficiency rather than emotional relief. Rational planning will normally require postponement of some tasks in favor of others to prevent gridlock and consequent paralysis of effort. Of course important tasks undone may generate their own worries, so the concerns of planning merge with those of mental health. A psychiatrist friend once gave me his criterion for a good administrator, namely the ability to tolerate a cluttered desk. His point, I take it, was that the urge to keep one's desk clean is likely to produce both inefficiency and mental stress.

Finally, there are certain tasks that themselves require a separation of phases. Writing is a prime example of such a task, where the first draft needs to be differentiated from the consequent editing phase. To collapse these two phases into one is a formula for paralysis. Some advisees of mine have begun work on their dissertations by attempting to frame a perfect first sentence. Finding this effort impossibly daunting, they have seen the completed dissertation receding more and more into the future, ever further away and beyond all hope of attainment. I have urged these advisees to put off editing until they had a sizable chunk to edit, asking them what they would think of a sculptor who attempted to create a human figure by first sculpting a perfect thumb. I wanted these students to procrastinate, to delay editing to a later time.

Even in creating a first draft, a writer needs to cut a large job into smaller bits, lest the conception of the whole intimidate him into silence. A student who says "Today I'm writing my dissertation" may find the day's task magnified beyond all feasible successful effort, whereas to say "Today I'm writing Section One of Chapter 1" marks out a doable job and liberates the writer from everything else. Procrastination here pays off.

A striking image, for me, of the two-phase task is the landing of a spaceship on the moon. About sixteen years ago, Joseph Weizenbaum of MIT, in a conversation with several friends at the Stanford Center, challenged us laymen to explain the remarkable fact that a rocket aimed at an object so far away could reach the target with unerring accuracy. He himself provided the answer. The initial shot required no pinpoint accuracy; it needed only to bring the spaceship into the general area of the moon, whereupon the ship's own guidance system could take over the residual task of putting the vehicle down. Procrastination respecting the actual landing made the whole task possible of execution.

## IDLENESS, OR LET THE WORLD GO BY

Idleness is surely the bane of American educators. Effort, diligence, activity—that's the ticket. From the behaviorist's industrial-efficiency model of schooling, through the progressives' learning-by-doing doctrine, to today's emphasis on "time-on-task," we have exalted the virtues of work/work/work in education. It is perhaps especially because American educational *practice* has been so relaxed, when compared with the practices of countries with more selective systems, that the emphasis on work has figured so strongly in American educational *theory*.

Now work is certainly a good thing, provided it is substantive and not mere busy work, provided intellectual advance is not confused with overt physical movements or mechanical drill or the shuffling of papers hither and yon. True intellectual advance is hard to track, however; it cannot be coerced or imposed as a duty, satisfying neither the moral puritan nor the efficiency expert with time clock at the ready, for it is often withdrawn and private, looking suspiciously like daydreaming. The silent pupil with the faraway look in his or her eyes may in fact be the one making the strongest mental gains.

Such a pupil was Albert Einstein, whose "aversion to the constant drill" in his gymnasium led to his being asked to leave the school (Frank, 1947). He had from early childhood "been inclined," in the words of one of his biographers, "to separate himself from children of his own age and to engage in daydreaming and meditative musing" (pp. 8, 10, 17). Of course, the young Albert was not your average student, but neither is it the case that only rare intellectual geniuses grow through exercise of the imagination. It must, in any event, give one pause to consider Einstein's mature criticism of what he calls coercive methods of education, requiring drill and cramming for set examinations:

> It is, in fact, nothing short of a miracle that the modern methods of instruction have not yet entirely strangled the holy curiosity of inquiry; for this delicate little plant, aside from stimulation, stands mainly in need of freedom; without this it goes to wreck and ruin without fail. It is a very grave mistake to think that the enjoyment of seeing and searching can be promoted by means of coercion and a sense of duty. To the contrary, I believe that it would be possible to rob even a healthy beast of prey of its voraciousness, if it were feasible, with the aid of a whip, to force the beast to devour continuously, even when not hungry, especially if the food, handed out under such coercion, were to be selected accordingly (see Einstein, 1949, pp. 17, 19).

Daydreaming may seem idle to the educational moralist — not work but mere play. We have then to remind the moralist that this form of play is often serious business and needs recognition and appreciation. In the words of John Dewey (1904, 1965), the "supreme mark and criterion of a teacher" is the ability to bypass externals and to "keep track of [the child's] mental play, to recognize the signs of its presence or absence, to know how it is initiated and maintained, how to test it by results attained, and to test *apparent* results by it" (p. 148).

Students of thinking tell us, further, that even real dreams may do intellectual work, and the history of science provides ample evidence. A striking case (Hempel, 1966) is Kekulé's discovery of the ring structure of the benzene molecule, after dozing in front of the fire and dreaming of snakes dancing in ringlike formations (p. 16). The notion of the incubation of a problem, where, without conscious attention to it, the mind silently works out a solution, has been long known and credited in experience. Sometimes the best approach to a problem is to turn away from it completely, let the mental machinery idle, go for a walk, take in a movie, have a cup of cocoa.

To acknowledge the role of what I have called "idleness" in education is to strive not only to address the pupil's mind and will, but also to capture his dreams, not only to assign him work but also to promote his play and to appreciate his spontaneity. This is a hard thing for teachers to do when they themselves are often viewed as minor technicians within a bureaucratic process, burdened by stringent social demands while their own spontaneity and initiative are unappreciated (see Scheffler, 1973; p. 61). The issue therefore has to do ultimately not just with teachers, but, as William James

(1958) recognized in his essay "The Gospel of Relaxation," with all of us who form the community within which education is carried on. His message is quite general: "*Unclamp* . . . your intellectual and practical machinery, and let it run free" (p. 144).

## CONCLUSION

I have now ended my argument. I trust you will not misunderstand me to be urging that our schools set about producing contrary, forgetful, idling ignoramuses, devoid of a proper sense of relevance and forever putting off until tomorrow what needs doing today. Rather, I have tried to rescue several virtues hidden in what are presently seen as unredeemable vices. Recognizing these virtues underneath their disguises can only serve, I am convinced, to enhance the processes by which we educate our children.

## REFERENCES

Black, M. (1967). Rules and Routines. In R. S. Peters (Ed.), *The concept of education* (pp. 92–104). London: Routledge & Kegan Paul.

Brown, S. I., & Walter, M. I. (1983). *The art of problem posing*. Philadelphia: The Franklin Institute Press.

Butler, J., D. C. L. (1900). *The analogy of religion: Natural and revealed, Vol. II of the works of Bishop Butler*. London: Macmillan.

Dewey, J. (1965). *The relation of theory to practice in the education of teachers*. (Reprinted in M. L. Borrowman, *Teacher education in America: A documentary history* [pp. 140–171]). NY: Teachers College Press.

Einstein, A. (1949). Autobiographical notes. P. A. Schilip (Tr.). In P. A. Schilip (Ed.), *Albert Einstein: Philosopher–scientist* (pp. 3–95). NY: Tudor.

Frank, P. (1947). *Einstein, his life and times*. NY: Knopf.

Hempel, C. G. (1966). *Philosophy of natural science*. Englewood Cliffs, NJ: Prentice Hall.

James, W. (1958). *Talks to teachers on psychology: And to students on some of life's ideals*. NY: Norton.

Luria, A. R. (1987). *The mind of the mnemonist*. Cambridge: Harvard University Press.

Mill, J. S. (1962). On Genius. In K. Price, *Education and philosophical thought* (pp. 540–547). Boston: Allyn & Bacon.

Oakeshott, M. (1966). Political education. In I. Scheffler (Ed.), *Philosophy and education*, (2nd ed., pp. 327–348). Boston: Allyn & Bacon.

Oakeshott, M. (1967). Learning and Teaching. In R. S. Peters (Ed.), *The concept of education* (pp. 156–176). London: Routledge & Kegan Paul.

Polanyi, M. (1964). *Science, faith and society*. Chicago: University of Chicago Press.

Popper, K. (1959). *The logic of scientific discovery*. London: Hutchinson.

Scheffler, I. (1965). *Conditions of Knowledge*, Chicago: University of Chicago Press.

Scheffler, I. (1973). *Reason and Teaching*. Indianapolis: Hackett.

Scheffler, I. (1986a). Computers at School. *Teachers College Record, 87,* 513–518.

Sward, K. (1948). *The legend of Henry Ford*. NY: Rinehart.

Walter, M. I. & Brown, S. I. (1969). What if not? *Mathematics Teaching, 46,* 38–45.

# II  Algebra and Arithmetic
## Editors' Comments

In this section we have grouped articles that deal mainly with arithmetic and algebra. The articles range from ones that can be used at a very early level in elementary school to ones that are appropriate even in college. They illustrate the wide variety of possibilities in the use of problem posing techniques.

These articles demonstrate that students get more involved in solving problems if they have posed them themselves. The articles indicate that students gain better insight into the mathematics they are learning when they vary different aspects of the phenomena they are studying. Some of the articles we have chosen emphasize standard curriculum topics while others focus on non-standard ones. In either case, they demonstrate the powerful insights one may acquire through the interplay of problem posing and problem solving.

While reading the articles, you might find it worthwhile to ask yourself with which aspects of the curriculum each article connects or which curriculum topics it amplifies. Alternately, you may wish to chose your favorite curriculum topic (which may mean not only the most fun to teach but the hardest to teach or motivate) and try your hand yourself at using some problem posing strategies to liven it up.

We have organized the essays in this Chapter into Four sections: **Asking Why, Mistakes, Tinkering With What Has Been Taken For Granted,** and **Your Turn.**

# Asking Why
# Editors' Comments

As we stated in **The Art of Problem Posing,** "it is worth asking *why* with many different intentions; that is, some *whys* call for calculations, some for insight into a gestalt, some for broader generalizations" (p. 113).

Whitin captures the power of a *why* question in his article **Number Sense and the Importance of Asking Why?** and shows that even young children can be encouraged to investigate their *why* questions. Here an eight-year-old child explores her *why* with the help of pictures. Whitin illustrates where a child can be led by encouraging her to pursue the *why* of "why does it not work now?" that the child posed while doing some arithmetical calculations. You might ask yourself how this *why* compares with those summarized above in the **Art of Problem Posing.**

Work such as described in this article can be extended and lead to later algebraic work. If $a + b = (a - d) + (b + d)$, why does $a \times b$ not equal $(a - d) \times (b + d)$? By how much is it off? Can it ever happen that $a \times b = (a - d) \times (b + d)$? Students may be amused when they find for example, that $32 \times 28 = (32 - 4) \times (28 + 4)$ and notice that this is an example of the commutative property in disguise. They can generalize and discover that one may add and subtract d from the factors, if and only if d is equal to the difference of a and b.

The same question that the student asks can occur in adult real life problems. Suppose one wants to find the cost of having a wooden floor scraped that is 33 feet × 17 feet. The cost of this depends on the square footage. Why does one not get the correct area of the floor by calculating $(33 - 3) \times (17 + 3)$ instead of $33 \times 17$? (see Walter, 1970).

We point out in **The Art of Problem Posing** that what at first may be a formal and rather stiff approach to using the What-If-Not technique (listing of attributes, followed by negating of attributes and listing of alternatives) gives way after some practice to carrying out the process in a less stiff way. Whitin's article indicates how listing and changing of attributes occurs naturally and less formally and can lead us to new questions and investigations.

Number tricks have often been used as a source for motivation but very often they are used only for providing motivation and practice in the four basic operations. Alternatively, they are used to entice students to investigate why a given trick works. Answering this *why* is certainly worthwhile and can often be illuminating. Among such tricks are ones such as: Write a three-digit number like 537 twice as in 537537. Such numbers are always divisible by 13. Small, in **Creating Number Problems,** Blake in **1089: An Example of Generating Problems** and Brutlag in **Making Your Own Rules** take number tricks as their starting point but turn them into a problem posing experience.

Both Small and Blake deal with posing problems starting with the '1089' number trick. In that trick, we take any three digit number, reverse the digits, subtract the smaller number from the larger number to get the difference, reverse the digits in that difference and add. The result is always 1089. Both authors apply the What-If-Not technique to the number trick. These two articles illustrate how different people can produce a different list of attributes starting with the same situation. Both articles generate rich new problems. Both authors ask also how the outcomes of various new cases are related. Knowing that the outcome of the trick in base ten is 1089, can the student predict what the outcome is in base six, for example? Blake ends his article with suggestions for different ways in which the What-If-Not technique can be used.

Small warns that both productive and unproductive problems may arise in using that technique and feels, as we do, that this situation is healthy for students to encounter. Textbooks tend to give problems that are always solvable and have an answer; real problem posing and solving is not like that. Often the problem is not even clear or if it is, it may be unmanageable unless more information is known or more restrictions are placed on what is given.

In **Making Your Own Rules,** Brutlag examines the number trick:

> Take any four digit number whose digits are not all the same. Rearrange the digits to form the largest number possible, that is, put the digits in decreasing order. Next, reverse the digits to form the smallest number possible and subtract the smaller number from the larger. Using the difference repeat the process.

Students are asked to try various starting numbers. Brutlag encourages his students to extend and modify the number trick using some of the What-If-Not techniques to do so. He encourages students to list and change the attributes of the starting number as well as of the operation involved in creating the new number. In this way his students get involved in making their own conjectures and in investigating them. The ideas of Brutlag's article can be applied to many other number tricks.

Brutlag shows why it is more manageable than one might predict for students to make up other rules for similar number 'tricks' by showing (using the pigeon hole principle) why one always will have to return to a number one has already encountered though it may take a long time to do so! Brutlag's article brought home to us that though students may be presented with number tricks they are rarely asked to think about how a trick was created in the first place or how new tricks might be discovered or how the students themselves can create one.

All three articles show ways students can get involved in making their own conjectures and in investigating them. These ideas can be applied to many other number tricks.

## REFERENCES

Walter, M. (1970). A common misconception about area. *The Arithmetic Teacher, 27,* 286–289.

# 11 Number Sense and the Importance of Asking "Why?"

David J. Whitin

When children ask the "why" question in their mathematics classroom, they are on their way to strengthening their number sense. An atmosphere that promotes curiosity about numbers encourages children to test their own hypotheses and pursue their own predictions. In this way children develop a broader understanding of why numbers act the way they do; they develop a number sense from the inside out because they are encouraged to be questioners and thinkers in their own right.

As an illustration, let us take the example of eight-year-old Tammy. She asked "why" and was encouraged to investigate further. Through her individual exploration she gained greater insight into mathematical properties and numerical patterns. Her story serves to highlight the importance of asking "why" in developing a greater number sense in children.

## TAMMY'S INVESTIGATION

As Tammy had worked with addition in her third-grade classroom, she had learned in an informal way about the compensation principle. She discovered that she could add a certain value to one addend and subtract that same value from the other addend and the sum would remain the same. For instance, $7 + 9$ was the same as $(7 - 1) + (9 + 1)$ or $(6 + 10)$. Thus, she saw that $7 + 9$ and $6 + 10$ both equaled the sum of 16 and could use blocks to prove it (fig. 11.1). She would often use this trick as she added. Later on during the school year, as Tammy started some beginning work with multiplication, she wondered whether this same mathematical principle could be applied to a new operation.

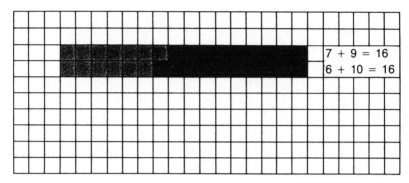

FIG. 11.1.

"Does 6 × 7 equal 5 × 8?" she asked.

The teacher, wanting Tammy to maintain interest in the problem, replied, "Try to find a way to figure it out."

After a short time Tammy returned with the results.

"It doesn't work," she said; "6 × 7 is 42 but 5 × 8 is only 40. It worked when you add numbers. *Why* doesn't it work now?"

Here was the "why" question that proved to be the catalyst for an interesting mathematical investigation. The source of this exploration was Tammy's original hypothesis – "I wonder if what I know about addition will help me with this new topic of multiplication." This question was the critical dimension of the experience. Without Tammy's willingness to take a risk and pursue a hunch, she never would have posed this "why" question. It is important that children experience a classroom atmosphere that encourages them to wonder about why numbers act the way they do and supports them in their need to discover the reasons.

The teacher gave Tammy the time to pursue her question by responding, "No, it does not work with these numbers. Try some other problems and see what you can find out. Can you find any numbers that do work?"

Tammy created several other problems. She applied the compensation strategy to the initial multiplication problem and compared the products. In no instance were the products the same:

$4 \times 6 = 24$                                  $5 \times 8 = 40$
$(4 - 1) \times (6 + 1)$                  $(5 + 1) \times (8 - 1)$
$3 \times 7 = 21$                                  $6 \times 7 = 42$


$9 \times 3 = 27$
$(9 - 1) \times (3 + 1)$
$8 \times 4 = 32$

Although all her examples proved that her addition trick did not work with multiplication, she still insisted on searching for an explanation. As she looked at her products she also wanted to know why "they got bigger or smaller by different amounts." Her further wondering about the behavior of numbers prompted Tammy's rich exploration of numerical relationships. Thus, asking an initial "why" question can often be a generative experience. One question engenders a host of others.

The teacher suggested that Tammy make some arrays on graph paper to illustrate her problems. Perhaps a visual representation could help to explain the answer to Tammy's question. She illustrated her first problem (fig. 11.2) by constructing a 4 × 6 array and a 3 × 7 array. The teacher suggested that she might want to superimpose the two arrays to see the change more clearly. Tammy did this as well. As she reviewed her drawings,

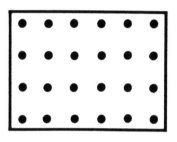

**Original array:**
4 rows by 6 columns = 24 elements

**Changed array:**
3 rows by 7 columns = 21 elements

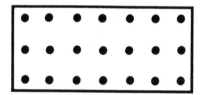

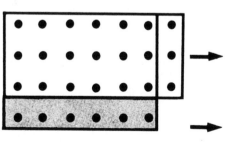

**Overlap:**
These 3 dots represent the number of elements added when the original array was changed. The gain is shown to be an array of three rows and one column (7 − 6), or 3 elements altogether.

This shaded area represents the number of elements lost when the original array was changed. This loss is shown to be an array of one row (4 − 3) by six columns, or six elements altogether.

Thus, the difference between the two arrays is 6 − 3, or 3.

FIG. 11.2.

Tammy observed, "There aren't the same number in a row and in a column. That's why it doesn't work." By decreasing the number of rows by one and increasing the number of columns by one, it was impossible to strike a balance. Exchanging one row for one column left an unequal amount. In the operation of addition, this exchanging, or compensation, worked successfully because the same amount was being added or subtracted each time. The same situation did not occur each time in multiplication, and the use of arrays helped to highlight this important difference for Tammy.

She also noticed that one way to ascertain the difference between the two arrays was to find the difference between the number of columns in one array and the number of rows in the other.

$$4 \times 6 = 24 \qquad 6 - 3 = 3$$
$$3 \times 7 = \underline{21} \qquad 7 - 4 = 3$$
$$3$$
difference

## ADDITIONAL QUESTIONS TO PURSUE

Tammy's investigation underscores the importance of allowing children the opportunity to pursue their own questions about the relative effects of operations on numbers. Teachers and children can generate further extensions by modifying some of the problem variables and posing such questions as these: What would happen if you did each of the following?

1. Increased one factor by two and decreased the other factor by two, for example, changed $7 \times 4 = 28$ to $5 \times 6 = 30$? Do your results produce a pattern? See figure 11.3.

2. Increased both factors by one, then two, then three? Is a pattern discernible?

$$3 \times 4 = 12 \quad 3 \times 4 = 12 \quad 3 \times 4 = 12$$
$$4 \times 5 = 20 \quad 5 \times 6 = 30 \quad 6 \times 7 = 42$$

3. Doubled one factor and halved the other factor?

$$5 \times 6 = 30$$
$$10 \times 3 = 30$$

or

$$5 \times 6 = 30$$
$$2.5 \times 12 = 30$$

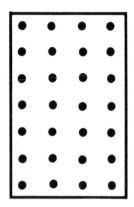

**Original array:**
7 rows by 4 columns = 28 elements

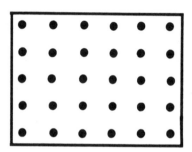

**Changed array:**
5 rows by 6 columns = 30 elements

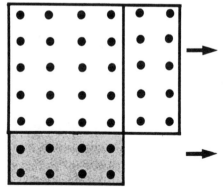

**Overlap:**
These 10 dots represent the number of elements added when the original array was changed. The gain is shown to be an array of five rows and two columns (6 − 4), or ten elements altogether.

This shaded area represents the number of elements lost when the original array was changed. This loss is shown to be an array of two rows (7 − 5) by four columns, or eight elements altogether. Thus, the difference between the two arrays is 10 − 8, or 2.

FIG. 11.3.  A schematic for question 1 of previous page.

Questions about operations other than multiplication could include these:

1. What would happen if you systematically —
   a) added one to one of the addends?

   $6 + 3 = 9$
   $7 + 3 = 10$
   $8 + 3 = 11$

   b) added one to each of the addends?

   $2 + 5 = 7$
   $3 + 6 = 9$
   $4 + 7 = 11$

   c) doubled both addends?

   $3 + 4 = 7$
   $6 + 8 = 14$

2. What would happen if you continued to double each set of addends? How does this change affect the sum?

   $12 + 16 = 28$
   $24 + 32 = 56$

3. What would happen if you systematically —
   a) increased the subtrahend of a subtraction equation by one?

   $11 - 4 = 7$
   $11 - 5 = 6$
   $11 - 6 = 5$

   b) decreased the minuend of a subtraction equation by one?

   $12 - 3 = 9$
   $11 - 3 = 8$
   $10 - 3 = 7$

   c) increased both the subtrahend and the minuend by one?

   $9 - 6 = 3$
   $10 - 7 = 3$
   $11 - 8 = 3$

4. What pattern emerges when you —
   a) double the dividend?

$$\begin{array}{r} 3 \\ 3\overline{)9} \end{array}$$

$$\begin{array}{r} 6 \\ 3\overline{)18} \end{array}$$

$$\begin{array}{r} 12 \\ 3\overline{)36} \end{array}$$

*b)* halve the divisor?

$$\frac{8}{12)\overline{96}}$$

$$\frac{16}{6)\overline{96}}$$

$$\frac{32}{3)\overline{96}}$$

*c)* double both the dividend and the divisor?

$$\frac{2}{5)\overline{10}}$$

$$\frac{2}{10)\overline{20}}$$

$$\frac{2}{20)\overline{40}}$$

The more opportunities children have to alter a particular problem or equation, the greater their understanding of that operation becomes. Number sense is strengthened when children are encouraged to "poke around" with numbers, gaining insight into the relative effects certain changes have on their solutions. This spirit of playfulness can highlight important numerical relationships and demonstrate the unique patterns that lurk beneath the surface as problem variables are systematically changed.

## LESSONS LEARNED

Tammy's investigation serves to document the many important reasons for encouraging children to ask "why" in the mathematics classroom. This one important question leads to the following benefits:

1. It strengthens a child's number sense. Tammy gained a broader understanding for the operations of addition and multiplication and an insight into the compensation law of arithmetic. Understanding how and why numbers work the way they do is an essential facet of one's "number sense."

2. It pushes learners beyond the right answer and forces them to explain the processes. The learner has no other alternative in pursuing the "why" question. Tammy's investigation helped to elucidate the process of multiplication and the law of compensation.

3. It fosters a spirit of inquisitiveness by conveying to learners the message that "I value how you think, not just what you know."

4. It empowers all participants in the classroom learning community. When pupils know that they can verbalize hunches, make predictions, and inquire about patterns, they are encouraged to take further risks. They know that their thinking will be respected. To stay within the parameters of present knowledge is not to grow. Only through pushing the boundaries of what they know can pupils deepen and broaden their number sense. In this kind of classroom environment no one knows the questions that will be asked because they are derived from the changing perceptions and experiences of the pupils. All feel strengthened in the knowledge that they will be given the support to make their own observations and pursue their own unique set of questions.

## THE CLASSROOM CLIMATE

It is essential that teachers create a classroom climate that encourages pupils to ask "why." The following are recommendations for designing such an environment:

1. *Follow a pupil's lead.* None of Tammy's investigations would have occurred if the teacher had not encouraged her to pursue her own question. Sometimes teachers claim they do not have the time to be "misled" by students' questions. Look again. Was Tammy misled? On the contrary, she was doing the leading, and in a most admirable fashion. When children gain number-sense abilities, they are encouraged by teachers who do more than just "cover" the material. These teachers take the time to "uncover" the material by listening to children and supporting them in their efforts to pose their own questions and devise their own solutions.

2. *Supply children with concrete materials.* Oftentimes the "why" question can be a rather abstract query, and it can be difficult to show the reasons that numbers act as they do. Having concrete materials available gives children the opportunity to pursue their question in a visual and concrete way. Materials help to furnish mental images for more abstract hypotheses. The size, color, and shape of various materials help to elucidate the mathematical processes.

3. *Ask questions about the process.* Often a single question can generate further inquiries as learners formulate additional hypotheses that they are eager to test. This questioning is the art of problem posing (Brown and Walter 1983); certain problem attributes are varied so that other dimensions of the problem can be pursued. The original problem becomes the seed for a multitude of related problems. In this way asking "why" spurs forward

teacher and student alike, guided by their own overarching spirit of mutual inquisitiveness.

## REFERENCE

Brown, S. I., & Walter, M. I. (1983). *The art of problem posing.* Hillsdale, NJ: Lawrence Erlbaum Associates.

# 12 Creating Number Problems

Marian Small

Marion Walter and Stephen Brown (1969) presented a problem posing technique which they called the "what if not" technique in an article in *Mathematics Teaching*. To employ this approach, the user lists each attribute of a situation, whether a theorem, a piece of equipment or a method of representation, and negates one attribute at a time to generate a new mathematical situation to be explored. Several other papers have been written by them and others (Brown & Walter, 1970; Walter & Brown, 1971; Schmidt, 1975) demonstrating the power of this method.

One of the standard number patterns we show children is to take any three-digit number, reverse the digits, subtract the smaller number from the larger to get a difference, reverse the digits in that difference and to add. Magically, we always get 1089, unless we start with a palindrome, in which case we always get 0. e.g.:

$$
\begin{array}{r}
542 \\
-245 \\
\hline
297
\end{array}
\qquad
\begin{array}{r}
297 \\
+792 \\
\hline
1089
\end{array}
$$

This is really a theorem with several attributes which we can list.

(i) Numbers are represented in a base ten numeration system.
(ii) Three-digit numerals are used.
(iii) Numerals are reversed.
(iv) Operations are addition and subtraction.
(v) Subtraction is performed before addition.

Students now consider what it would mean to negate these attributes. The negations suggested by my University class were:

(i) Work in whole number bases other than base ten.
(ii) Use two-, four- or five-digit numerals.
(iii) Reverse just the first and last digits of the original numeral.
(iv) Operate with multiplication and division.
(v) Add first and then subtract.

Some of these investigations turned out to be productive and others less so. It is important for students to see that mathematics is not always so clear-cut that every problem has a neat and simple solution.

Let us examine what happens if we work in a base other than base ten. We started with base nine.

| | | |
|---|---|---|
| 432 | 523 | 411 |
| − 234 | − 325 | − 114 |
| 187 | 187 | 286 |
| + 781 | + 781 | + 682 |
| 1078 | 1078 | 1078 |

The theorem was that in base nine the result of the procedure is 1078. This was easily proved. We then guessed a theorem for base eight of 1067, for base seven of 1056, for base six of 1045, for base five of 1034, for base four of 1023, for base three of 1012, and for base two of 1001. Any of these can be verified. We extended to the direction of bases greater than ten as well. In base $b$, the result is $10(b - 2) (b - 1)$.

A more encompassing investigation is based on negation (ii). What happens if we use two-, four- or five-digit numerals? We found that with two-digit numerals, our result was 99 for non-palindromes, but with four-digit numerals, the results became complicated. We obtained sums of either 0, 990, 9999, 10890 or 10989, depending on certain conditions relating the digits $a$, $b$, $c$ and $d$ of the numeral '$abcd$'. If $a = d$ and $b = c$ (a palindrome), we got 0 as before. If $a = d$ and $b > c$, we got 990 (the entire four-digit difference is reversed, including leading zeros). If $a > d$ and $b \leq c - 1$, we got 9999. If $a > d$ and $b \geq c + 1$, we got 10890. If $a > d$ and $b = c$, we got 10989.

First of all it was curious that our simple theorem became so complicated. One student observed that in this case the middle subtraction was not as predetermined as in the three-digit case. This led me to explore a situation where the middle subtraction is more like that in the three-digit case; this is achieved by using negation (iii) along with negation (ii), reversing just the first and last digits of the numeral. In a two or three digit problem, this is equivalent to reversing the entire numeral, but not in the larger digit problem. We also observed with some interest that even though the situation is muddled when the numeral is completely reversed, the important digits are still 1, 0, 9 and 8; this led to a discussion of why this might be so. I checked to see whether this still occurred with five-digit numerals (it does!) and whether in base nine, for example, the important digits are 1, 0, 8 and 7 (they are!).

I pursued negation (iii) combined with negation (ii), and found that with four-digit numerals, the result of reversing, subtracting, reversing and adding is 10989, with five digits it is 109989, with six digits it is 1099989, etc. This condition of reversing only the first and last digits is, indeed, an intrinsic part of the original pattern.

In trying to use other operations than addition and subtraction, negation (iv), my students ran into greater difficulties. The problem, of course, is that in dividing we are forced to use decimals and we do not really know how to reverse the digits. Take, for example, a two-digit problem. Start with 41; reverse to 14. Multiply $41 \times 14 = 574$. Reverse to 475. Then take $574 \div 475$. It is a mess! The student encounters his first real barrier. He learns to think about the utility of pursuing a certain problem and begins to exercise mathematical judgment.

Pursuing negation (v) and altering the order of operations is slightly more productive, but still not altogether clear cut. We took some sample three-digit numerals, reversed the digits and added. We reversed the digits in our answer and subtracted.

|  |  |  |
|---|---|---|
| 523 | 582 | 423 |
| +325 | +285 | +324 |
| 848 | 867 | 747 |
| −848 | −768 | −747 |
| 0 | 99 | 0 |

|  |  |
|---|---|
| 471 | 723 |
| +174 | +327 |
| 645 | 1050 |
| −546 | −0501 |
| 99 | 449 |

It looked good until the last example where we ran into trouble since our first answer (sum) was four digits, rather than three. It appears that if we represent our number as '$abc$', then when $a + c < 10$, our result is 0 or 99 depending on whether $b < 5$. Otherwise, we have a situation where the result heavily depends on the initial number. We wondered why the four-digit situation is a problem. By representing the situation algebraically, students saw that the numbers which are subtracted within a given column do not relate to each other as closely as in the case of the three-digit sum.

Both productive and 'unproductive' problems arise from the basic investigation. This is a healthy situation for mathematics students. The number patterns are sufficiently interesting, I think, to sustain the student's motivation to explore the problems generated. Moreover, the student learns how mathematics is created and may gain some ability to create some of his own theorems.

## REFERENCES

Brown, S. I., & Walter, M. I. (1970). What if not? An elaboration and second illustration. *Mathematics Teaching, 51,* 9–17.

Schmidt, P. J. (1975). A non-simply connected geo-board: Based on the what-if-not-idea. *Mathematics Teacher, 68,* 384–388.

Walter, M. I., & Brown, S. I. (1969). What if not? *Mathematics Teaching, 46,* 38–45.

Walter, M. I., & Brown, S. I. (1971). Missing ingredients in teacher training: One remedy. *American Mathematical Monthly, 78,* 399–404.

# 13 Making Your Own Rules

Dan Brutlag

"Making conjectures, gathering evidence, and building an argument to support such notions is fundamental to doing mathematics" (NCTM 1989, 7). This statement represents one of the five curricular goals found in the introduction to NCTM's *Curriculum and Evaluation Standards for School Mathematics*. However, the work that students typically do in mathematics classes consists of proving someone else's conjectures, gathering evidence according to someone else's directions, and imitating someone else's arguments. This article presents a method I have used to help students start their own original investigations and create their own conjectures.

## INVESTIGATIONS

Our own investigations arose from the "6174 problem" in our textbook (Bennett and Nelson 1979, 129). The rule given for the 6174 problem is as follows:

> Take any four digit number whose digits are not all the same. Rearrange the digits to form the largest number possible, that is, put the digits in decreasing order. Next, reverse the digits to form the smallest number possible and subtract the smaller number from the larger. Using the difference, repeat the process.

What happens if the process is followed over and over? For example, suppose we start with 7992. If we rearrange the digits according to the rule,

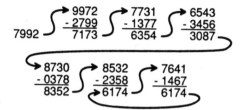

FIG. 13.1.    The "6174 problem"

we get 9972. Then 9972 − 2799 = 7173. Figure 13.1 shows what happens when the rule is applied over and over. The sequence generated by the rule is

7992, 7173, 6354, 3087,
8352, 6174, 6174, 6174, . . . .

Note that since 6174 was obtained from 6174, it would also be the result on the seventh and all succeeding applications. The magic in the 6174 problem is that no matter what starting number is used, after a few applications of the rule the sequence always cycles on 6174.

I regularly encourage my students to extend and modify problems. One of my students decided to try the rule of the 6174 problem on three-digit numbers. Her sequences led to 495 every time! I had already been thinking about the 6174 problem myself and had conjectured that if *any* rule was applied over and over for different starting numbers, the sequence of numbers generated would *always* lead to a particular number or numbers, or form some repeating number pattern. I decided to have my class help test my conjecture.

## "MAKE YOUR OWN RULES" ASSIGNMENT

I gave the class the following assignment:

> Make up a set of rules that can be applied to any three-digit number and that gives another number with three or fewer digits. Apply your rules over and over to different three-digit starting numbers. See if anything interesting happens.

To help students devise their own investigations we first generated two lists, a "number-operations list" and a "number-properties list." The method of generating lists of attributes about a particular problem or situation to make up new problems is described in the book *The Art of Problem Posing* (Brown and Walter 1983). The lists served as a mental jump-start for students to begin the class assignment (table 13.1).

TABLE 13.1
Attribute Lists

| Number-Operations List | Number-Properties List |
|---|---|
| Add ____ to ____ | Divisible by ____ |
| Move to another position | Hundreds, tens, or |
| Take positive difference | units place |
| Double | Prime |
| Square | Composite |
| Halve if even | Even, odd |
| Multiply by ____ | Multiple of ____ |
| Replace by ____ | Greater than |
| Exchange | Less than |

The number-operations list could be used to make up such instructions as "Replace the hundreds and tens digits with double their sum." The lists could also be used together to make up such conditional statements as "If the number is composite, then divide it by its greatest factor less than itself."

## SOME STUDENT RULES AND CONJECTURES

Here are some of the rules and conjectures originated by students in my classes:

### Lori's rule

Start with any three-digit number. To get the hundreds digit of the next number in the sequence, take the starting number's hundreds digit and double it. If the double is more than 9 then add the double's digits together to get a one-digit number. Do the same thing to the tens and units digits of the starting number to obtain the tens and units digits of the new number.

Example, starting with 567:

567, 135, 261, 432, 864,
738, 567, 135, 261, . . .

*Conjecture.* After exactly six steps, you will always get your original number again.

### Barbette's rule

Start with any three-digit number. Add all the digits together and multiply by 2 to get the hundreds and tens digits of the next number in the sequence.

To get the units digit of the next number, take the starting number's tens and units digits and add them. If the sum is more than 9, add the digits again to get a one-digit number for the units place.
Example, starting with 563:

563, 289, 388, 387, 366, 303, 123,
125, 167, 284, 283, 262, 208, . . .

*Conjecture.* No matter what number you start with, 208 will always occur in the sequence of numbers generated.

## Ramona's rule

Start with any three-digit number. Obtain the next number in the sequence by moving the hundreds digit of the starting number into the tens place of the next number, the tens digit of the original number into the units place of the next number, and the units digit into the hundreds place. Then add 2 to the units digit of the new number; however, if the sum is greater than 9 use only the units digit of the sum.
Example, starting with 324:

324, 434, 445, 546, 656, 667, 768, 878,
889, 980, 90, 1, 102, 212, 223, 324, . . .

*Conjecture.* You will always obtain your original number after exactly fifteen steps.

## Lisa's rule

Start with any three-digit number. If the number is a multiple of 3 then divide it by 3 to get a new number. If it is not a multiple of 3, then get a new number by squaring the sum of the digits of the number.
Example, starting with 315:

315, 105, 35, 64, 100, 1, 1, 1, . . .

Starting with 723:

723, 241, 49, 169, 256, 169, 256, . . .

*Conjecture.* No matter what starting number is chosen, after enough applications of the rule the sequence will always go to 1, 1, 1, . . . or to the cycle 169, 256, 169, 256, . . . .

## DETAILS OF LISA'S INVESTIGATION

The investigation of Lisa's rule begins by looking at some number sequences generated by different starting numbers. Here are a few starting numbers and the sequences that follow each:

834, 278, 289, 361, 100, 1, 1, 1, . . .
125, 64, 100, 1, 1, 1, . . .
567, 189, 63, 21, 7, 49, 169, 256, 169, 256, . . .
643, 169, 256, 169, 256, . . .
365, 196, 256, 169, 256, 169, 256, . . .

Students who have knowledge of computer programming can use the computer to do the repetitive calculations quickly. Program 1 will print out the results of applying Lisa's rule over and over for any three-digit starting number.

Looking at many sequences gives insight as to why Lisa's rule always gives sequences that lead to

1, 1, 1, . . .

or to

169, 256, 169, 256, . . .

Suppose Lisa's rule is to be used on a starting number that has 3 as a factor $k$ times. Then after $k$ applications of the rule, the resulting number will not be a multiple of 3, since all the factors of 3 will have been divided out. Thus all starting numbers, either immediately or after a few applications of the rule, will generate sequences that lead to a non-multiple of 3.

---
PROGRAM 1
---

```
10   INPUT " TYPE A THREE-DIGIT NUMBER:
         " ; N
20   IF INT(N/3)><N/3 THEN 50
30   LET N = N/3
40   GO TO 90
50   LET A = INT(N/100)
60   LET B = INT(N/10) - 10*A
70   LET C = N - (100*A + 10*B)
80   LET N + (A + B + C) ^ 2
90   PRINT N; "    ";
100   INPUT " ANOTHER? (Y OR N): " ; Z$
110   IF Z$ = " Y " THEN 20
120   END
```

Once a sequence reaches a non-multiple of 3, all numbers in the sequence that follow will have two true conditions:

1. the sum of the number's digits will be non-multiples of 3, and
2. the maximum value for the sum of digits will be 26 and the minimum will be 1, since the largest possible non-multiple of 3 is 998.

Therefore, no matter what the starting number, every sequence after a few applications of the rule will get to one of the squares of the numbers from 1 to 26 that are not multiples of 3, that is, the squares of 1, 2, 4, 5, 7, . . ., 23, 25, 26. Examining the sequences obtained by starting with each of these numbers shows that each leads either to 1 or to the cycle 169, 256.

## THE PIGEONHOLE PRINCIPLE

My original conjecture that prompted this assignment was that *any* rule applied over and over for different starting numbers would generate a sequence of numbers that would *always* either lead to a particular number or numbers, or form some repeating number pattern. My students have explored many different rules, and so far the magic has always happened — interesting number sequences have always occurred.

One underlying mathematical concept that applies is the "pigeonhole principle." Suppose we have six pigeons that are to be placed into five pigeonholes. Applying the pigeonhole principle to this situation leads to the conclusion that *at least one of the pigeonholes must contain two or more pigeons* (Long 1988).

Recall that the assignment was to make up rules that result in three- or fewer-digit numbers. Since only 1000 integral numbers occur between 0 and 999, inclusive, the pigeonhole principle guarantees that sooner or later the rule will generate a number that has previously appeared in the sequence. At this point the sequence will cycle. At worst, a cycle will occur after the one-thousandth application of a rule. Fortunately, almost every rule we have investigated has led to something interesting much sooner than the thousandth application!

## CONCLUSION

Unexpected number patterns fascinate and surprise students, especially when the patterns result from the students' own investigations. Some of my students' conjectures go beyond their abilities to prove or disprove; but proving or disproving conjectures is not the main goal of this activity. In doing mathematics, a deductive proof is usually the last thing done in an investigation. For mathematicians, the deductive argument is the frosting on the cake. But for many students, a deductive argument, given without any personal investigation, is the last nail in the coffin. Students need more inductive mathematical experiences to help them see mathematics as a creative, dynamic subject.

## REFERENCES

Bennett, A. B., & Nelson, L. T. (1979). *Mathematics: An informal approach.* Newton, MA: Allyn & Bacon.

Brown, S. I., & Walter, M. I. (1983). *The art of problem posing.* Hillsdale, NJ: Lawrence Erlbaum Associates.

Long, C. T. (1988). On pigeons and problems. *Mathematics Teacher, 81,* 28–30, 64.

National Council of Teachers of Mathematics. (1989). *Curriculum and evaluation standards for school mathematics.* Reston VA: Author.

# 14 1089: An Example of Generating Problems

Rick N. Blake

Emphasis on problem solving in mathematics has gained considerable attention in the last few years. A joint position paper on basic mathematical skills by the National Council of Teachers of Mathematics and the National Council of Supervisors of Mathematics, in the February 1978 *Mathematics Teacher,* stated that "learning to solve problems is the principal reason for studying mathematics." More recently the NCTM, in its booklet entitled *An Agenda for Action: Recommendations for School Mathematics of the 1980s,* strongly recommends that "problem solving be the focus of school mathematics in the 1980s."

For problem solving to be the central theme of school mathematics, we will need to provide our students with both a knowledge of the procedures used to solve problems and practice in solving problems. Yet where can we obtain problems? One way is to generate them, or better yet — to involve our students in generating and solving their own problems. The purpose of this article is to illustrate a method for generating new problems by using a number puzzle.

## THE 1089 PUZZLE

Start with a three-digit number in which the hundreds digit is at least two more than the ones digit. Reverse the digits and subtract the smaller number from the larger to obtain a difference. Now, reverse the digits of the difference and add.

141

"Magically," we should always obtain 1089. For example, using the three-digit number 723 yields the following:

$$
\begin{array}{r} 723 \\ -327 \\ \hline 396 \end{array}
\quad \text{and} \quad
\begin{array}{r} 396 \\ +693 \\ \hline 1089 \end{array}
$$

Let us look briefly at two arguments to show that 1089 will always be the solution.

## An Algebraic Argument

Let $abc$ be our three-digit number, with $a \geq c + 2$. Now, reverse the digits and subtract (Fig. 14.1).

$$
\begin{array}{ccc}
a & b & c \\
-c & b & a \\
\hline
(a - 1 - c) & (10 + b - 1 - b) & (10 + c - a)
\end{array}
$$

FIG. 14.1.

Recall that $a > c$, so we must borrow, and for $a - 1 - c$ not to be 0, we need $a > c + 2$. The tens digit simplifies to 9. Now, reverse the digits of the difference and add (Fig. 14.2).

$$
\begin{array}{ccc}
(a - 1 - c) & 9 & (10 + c - a) \\
+ (10 + c - a) & 9 & (a - 1 - c) \\
\hline
(a - 1 - c + 10 + c - a) & (18) & (10 + c - a + a - 1 - c)
\end{array}
$$

FIG. 14.2.

This sum simplifies to 1089.

## An Inductive Argument

If we look at some examples and note the differences between the hundreds and ones digits, $a$ and $c$, we can generate Table 14.1. What patterns can we see? The tens digit is always 9, and the sum of the hundreds digit and the ones digit is always 9. So our difference after subtracting is $x9y$, where $x + y = 9$. If we reverse these digits and add, we obtain the following:

$$
\begin{array}{r} x9y \\ y9x \\ \hline 1089 \end{array}
$$

TABLE 14.1.
Data for Inductive Argument

| Difference between a and c | Difference after Subtracting | Reversal of Difference |
|---|---|---|
| 2 | 198 | 891 |
| 3 | 297 | 792 |
| 4 | 396 | 693 |
| 5 | 495 | 594 |
| 6 | 594 | 495 |
| 7 | 693 | 396 |
| 8 | 792 | 297 |
| 9 | 891 | 198 |

## THE "WHAT-IF-NOT" TECHNIQUE

The "what-if-not" technique was developed some years ago by Stephen Brown and Marion Walter while they were team teaching a graduate course in problem solving at Harvard University. To employ the procedure, we identify the attributes of a theorem, problem, or puzzle and negate them one at a time. Each negation modifies the original problem and offers the potential for generating new problems.

To apply this technique to the 1089 problem, we first must identify the attributes of the problem.

1. Three-digit numbers are used.
2. The hundreds digit is at least 2 greater than the ones digit.
3. The numbers are represented in base ten.
4. The digits are reversed.
5. The operations are subtraction, then addition.

The second stage of the technique is to negate these attributes one at a time and then, based on each negation, attempt to generate new solutions. In some cases we may not be able to pose a new problem, or perhaps many problems could be posed. Keeping in mind that our original situation is a numerical puzzle with a constant outcome, we might want to consider two questions as we look at each new situation: (1) Can we predict the outcome? (2) How, if at all, are the outcomes related to each other?

### What If We Do Not Use Three-Digit Numbers?

This circumstance involves a number of new problems. Suppose we start with a two-digit number (Table 14.2). In example (iii), the tens digit is only one larger than the units digit, but if we keep the condition that the outside

TABLE 14.2.
Some Examples with Two Digits

| | | | | |
|---|---|---|---|---|
| (i) | 72 | and | 45 | |
| | − 27 | | + 54 | |
| | 45 | | 99 | |
| (ii) | 81 | and | 63 | |
| | − 18 | | + 36 | |
| | 63 | | 99 | |
| (iii) | 54 | and | 9 | |
| | − 45 | | + 9 | |
| | 9 | | 18 | |
| (iv) | 54 | and | 09 | |
| | − 45 | | + 90 | |
| | 09 | | 99 | |

digits differ by at least 2, we will always have a solution of 99, or if we reverse the difference, of 09 to 90, as in (iv), we obtain 99. If we use a four-digit number, say *abcd,* we obtain sums of 0, 990, 10989, 10890, or 9999, depending on the relationship among the original four digits.

If $a = d$ and $b = c$ (a palindrome), we obtain 0.
If $a = d$ and $b > c$, we obtain 990 (if we reverse the entire four-digit number including the 0).
If $a > d$ and $b = c$, we obtain 10989.
If $a > d$ and $b > c$, we obtain 10890.
If $a > d$ and $b < c$, we obtain 9999.

All in all, the solution is not very "nice." In fact, if we examine our work, we will find no predictable pattern after the first subtraction (as we found in the three-digit case where the tens digit was always 9 and the hundreds and ones digits added to 9) except where $b$ and $c$ are equal.

This process might lead us to examine what we mean by reversing the digits. In the three-digit example, *abc,* when we reversed the digits, *cba,* the middle digit remained in the same place. If we reverse the digits of a four-digit number, *abcd,* where $b = c$, we obtain *dbca,* that is, we only switch the two outside digits. If we define *reverse the digits* to mean interchange the two outside digits, what will our result be?

| | | |
|---|---|---|
| 5892 | | 2997 |
| − 2895 | and | + 7992 |
| 2997 | | 10989 |

Notice, using this definition, that we obtain a solution of 10989, and we also have a predictable pattern after the first difference. This pattern is similar to that for a three-digit number. If we extend the number of digits to five and use our new definition of reversing the digits, we obtain 109989, as in this example:

$$
\begin{array}{r}
57431 \\
-17435 \\
\hline
39996
\end{array}
\qquad \text{and} \qquad
\begin{array}{r}
39996 \\
+69993 \\
\hline
109989
\end{array}
$$

These patterns will extend to any number of digits. We summarize our findings in Table 14.3.

From the Table 14.3 we can find a number of interesting patterns.

- $x + y = 9$, and $x$ is 1 less than the difference between the two outside digits.
- The number of 9s in both the difference and the sum increase as the number of digits increases, and we can predict the number of times they occur.
- Two interesting relationships occur between the sums, one involving 99 as a factor and the other involving 9 as a factor.

## What If the Hundreds Digit is Not at Least Two More Than the Ones Digit?

Here again, there are a number of examples to consider. The results are as follows:

If $a = c$ (a palindrome), we obtain 0.
If $a = c - 1$, we obtain $- 198$ or $- 1089$.
If $a \leq c - 2$, we obtain $- 1089$.

## What If the Numbers are Not in Base Ten?

Suppose we use six for a base:

$$
\begin{array}{r}
412 \\
-214 \\
\hline
154
\end{array}
\qquad \text{and} \qquad
\begin{array}{r}
154 \\
+451 \\
\hline
1045
\end{array}
$$

or

$$
\begin{array}{r}
501 \\
-105 \\
\hline
352
\end{array}
\qquad \text{and} \qquad
\begin{array}{r}
352 \\
+253 \\
\hline
1045
\end{array}
$$

Notice a similar pattern in the first difference: the middle digit is 5 (the largest digit in base six), and the sum of the two outside digits is also 5.

**TABLE 14.3**
**Vary Number of Digits**

| No. of Digits | Difference after Subtracting ($x + y = 9$) | Sum (Answer) | Sum Factored |
|---|---|---|---|
| 2 | $xy$ | 99 | $1 \cdot 99 = 9 \cdot 11$ |
| 3 | $x9y$ | 1089 | $11 \cdot 99 = 9 \cdot 121$ |
| 4 | $x99y$ | 10989 | $111 \cdot 99 = 9 \cdot 1221$ |
| 5 | $x999y$ | 109989 | $1111 \cdot 99 = 9 \cdot 12221$ |
| 6 | $x9999y$ | 1088889 | $11111 \cdot 99 = 9 \cdot 122221$ |
| ... | | | |
| 15 | $\left. \begin{array}{c} 13 \\ \overbrace{\phantom{xx}} \end{array} \right.$ $x99\ldots9y$ | $\left. \begin{array}{c} 12 \\ \overbrace{\phantom{xx}} \end{array} \right.$ $1099\ldots989$ | $\left. \begin{array}{c} 14 \\ \overbrace{\phantom{xx}} \end{array} \right.$ $11\ldots1 \cdot 99 = 9 \cdot 122\ldots21$ $\left. \begin{array}{c} 13 \\ \underbrace{\phantom{xx}} \end{array} \right.$ |

If $a \geq c + 2$ (original), we obtain 1089.

If $a = c + 1$, we obtain 198, or if we reverse the entire number including the 0, 1089.

If we use 9 as a base, we have the following:

$$
\begin{array}{r}
412 \\
-214 \\
\hline
187
\end{array}
\qquad \text{and} \qquad
\begin{array}{r}
187 \\
+781 \\
\hline
1078
\end{array}
$$

or

$$
\begin{array}{r}
835 \\
-538 \\
\hline
286
\end{array}
\qquad \text{and} \qquad
\begin{array}{r}
286 \\
+682 \\
\hline
1078
\end{array}
$$

Can we predict the first difference and the sum in any base? The answer is — of course we can!

If the base is $b$, then the first difference will be $x(b - 1)y$, where $x + y = b - 1$ and $x$ is one less than the difference of the two outside digits of the original number, and the final result will be

$$10(b - 2)(b - 1).$$

## What If We Do Not Reverse the Ones and Hundreds Digits?

One of the problems we can consider is this: what if we reverse the digits and subtract, but do not reverse them when we add?

$$
\begin{array}{r}
542 \\
-245 \\
\hline
297
\end{array}
\qquad \text{and} \qquad
\begin{array}{r}
297 \\
+297 \\
\hline
594
\end{array}
$$

or

$$
\begin{array}{r}
412 \\
-214 \\
\hline
198
\end{array}
\qquad \text{and} \qquad
\begin{array}{r}
198 \\
+198 \\
\hline
396
\end{array}
$$

Again, we see some familiar patterns. Our pattern for the first difference still remains, but this time our sum, although not constant, exhibits a pattern similar to the first difference patterns. If we organize the data in Table 14.4, we can look for some patterns.

## TABLE 14.4
### Reverse Only Original Digits

| Difference of Ones and Hundreds Digits | Difference | Sum | Sum Factored |
|---|---|---|---|
| 1 | 99 | 198 | 2·1·9·11 = 1·198 |
| 2 | 198 | 396 | 2·2·9·11 = 2·198 |
| 3 | 297 | 594 | 2·3·9·11 = 3·198 |
| 4 | 396 | 792 | 2·4·9·11 = 4·198 |
| 5 | 495 | 990 | 2·5·9·11 = 5·198 |
| 6 | 594 | 1188 | 2·6·9·11 = 6·198 |
| 7 | 693 | 1386 | 2·7·9·11 = 7·198 |
| 8 | 792 | 1584 | 2·8·9·11 = 8·198 |
| 9 | 891 | 1782 | 2·9·9·11 = 9·198 |

The patterns in the sum can be described in various ways. If $x$ is the difference between the ones and hundreds digits, then —

- the sum is $x \cdot 198$;
- the number of hundreds is the $x$th odd number $(2x - 1)$, and the sum of alternate digits is 9, that is, if the sum is $abcd$, then $a + c = b + d = 9$.

Here are some other problems we could consider as well:

- What if we reverse the hundreds and tens digits?
- What if we reverse the tens and ones digits?
- What if we reverse just the digits of the difference?
- What if we apply the what-if-not technique to this problem? Can we predict the outcome if we use a different base or a different number of digits?

## What If We Do Not Subtract First, and Then Add?

This condition produces a number of problems that could be investigated.

- What if we do not add, but subtract twice instead?
- What if we add twice?
- What if we add first, and then subtract?
- What if we use other operations, that is, multiplication or division?

What have we done up to this point? We have generated a lot of problems (not all) using one numerical puzzle. We have attempted to solve some of these problems by collecting data, organizing it (in most cases into a table), and looking for patterns.

We have by no means exhausted the questions one might raise; we could explore still many other questions related to this puzzle. We could also apply the what-if-not technique to any of the new problems we have generated, or to groups of them. It may be that if we changed one of the operations or the number of digits or the base, we could find some very interesting relationships.

The example used in this article is from number theory and lends itself to working with patterns. However, the what-if-not technique can be applied to many problems, theorems, or puzzles. For example, many problems dealing with gear ratio, measuring time, probability, or kite flying lend themselves to this technique. Additional examples from number theory and geometry can be found in the Bibliography.

## SOME IDEAS FOR USING THE "WHAT-IF-NOT" TECHNIQUE

1. It can generate problems to be assigned or used in class. These problems could follow directly from using the what-if-not procedure or from patterns and relationships that you have found; for example:

Do the 1089 puzzles in base six and base eight. Can you predict what the final result will be using these bases? Can you find a relationship between the base used and the final result for any base? Using your calculator to do all computations, find

$$109999999989 \div 99.$$

You may use paper and pencil for recording only.

2. Use the technique in class to generate and then to solve problems — essentially what we have done in this article. A problem can be presented and looked at in class. Its attributes can be identified and negated. Students can work on generating and solving related problems either individually, in small groups (perhaps with each group looking at a different attribute), or as a whole class.

3. Use the technique on homework assignments to have students generate problems and solve them. List the attributes of the problem and have the student negate them one at a time; generate related problems and investigate them.

4. Use the procedure as an investigation, in class or as homework. For this use you would state the problem and ask the students to apply the what-if-not procedure and to generate and solve or discuss as many problems as they can.

It should be kept in mind that at times problems will be generated by this procedure that either do not have a solution or whose solution is beyond the ability of your students. How open-endedly you use the technique depends on the ability and background of your students.

We need to get our students involved in the problem-solving process — solving problems, generating problems, and looking at relationships between problems. Many of the heuristics we use in problem solving focus on these ideas:

- Can I solve a simpler problem?
- Can I solve a more general problem?
- Have I solved a problem before whose solution I can use?
- Have I solved a problem before whose solution process I can use?
- Can I solve part of this problem?
- Can I forget about part of the information given and solve the problem?
- Can I work the problem backward?

In all these questions we are generating new situations or looking at relationships among them. The what-if-not technique is a process that can get students involved in generating problems and in looking for relationships among problems.

## REFERENCES

Brown, S. (1976). From the golden rectangle and fibonacci to pedagogy and problem posing. *Mathematics Teacher, 69,* 180–88.

Brown, S. (1981). "Ye Shall Be Known by Your Generations." *For the Learning of Mathematics*, 3, 27–36.

Brown, S. & Walter, M. (1970). What if not? An elaboration and second illustration. *Mathematics Teaching, 51,* 9–17.

Brown, S., & Walter, M. (1972). "The Roles of the Specific and General Cases in Problem Posing." *Mathematics Teaching, 59,* 52–54.

Schmidt, P. (1975). A Non-simply connected geoboard based on the 'what if not' idea. *Mathematics Teacher 68,* 384–388.

Small, M. (1977). Creating number problems. *Mathematics Teaching, 81,* 42–43.

Walter, M. & Brown, S. (1969). What If Not? *Mathematics Teaching, 46,* 38–45.

Walter, M. I., & Brown, S. I. (1971). "Missing Ingredients in Teacher Training: One Remedy." *American Mathematical Monthly,* 78, 399–404.

Walter, M. I., & Brown, S. I. (1977). "Problem Posing and Problem Solving: An Illustration of Their Interdependence." *Mathematics Teacher, 70,* 4–13.

# Mistakes
# Editors' Comments

As teachers we too often dismiss a "mistake" as something to be avoided and at best to be corrected as soon as possible. Meyerson in **Mathematics Mistakes** looks at many types of algebraic mistakes and some purely arithmetic ones and discusses how they can be channeled into a positive, useful, learning and growing experience. He suggests that they can be used as a springboard for developing new mathematics and can be used to turn negative experiences into positive ones.

Borasi in **Algebraic Explorations of the Error 16/64 = 1/4** examines in detail one common arithmetic mistake in which the right answer is obtained by an incorrect procedure and shows how such errors can lead to problem posing as well as problem solving activities. Her techniques can be used at a variety of levels and applied to other "mistakes."

# 15 Mathematical Mistakes

Lawrence Nils Meyerson

Too much of what is taught in the schools comes directly from the classroom teacher or the textbook and not enough as a *reaction* to pupil perceptions. We often glide through a lesson with satisfaction as long as our pupils are feeding back what we want them to learn. When they make a mistake we simply correct it and forge ahead. Very rarely do we capitalise on a mistake as an incident of high potential. Many teachers would automatically categorise a mistake as an evaluative experience for both the teacher and the pupil, but this is not always a necessary avenue to follow. There are other options.

The following is a probable treatment of a classical mathematical mistake. The pupil squares $a + b$ to get $a^2 + b^2$, leaving out the $2ab$ term.

1. The teacher may ask the pupil to substitute a numerical example; comparing that result with normal multiplication will hopefully convince the student that

$$(a + b)^2 \neq a^2 + b^2.$$

2. He may show the correct method of squaring $a + b$, algebraically or geometrically, thus introducing the 'middle term' as something demonstrated by proof. The first option uses falsifiability to persuade the pupil; the second provides 'positive' proof.

This system of showing the pupil that his belief is wrong and then countering with the correct method makes two basic assumptions:

(i) the pupil's belief in his original 'mistake' was not strong;
(ii) the error was roughly speaking 'random'; that is, the pupil did not base his conclusion on any mathematical or psychological idea previously explored by the teacher or the pupil.

There are cases where these assumptions do hold, but is substituting numerical examples enough to dispel a *very strong* belief that

$$(a + b)^2 = a^2 + b^2,$$

and more importantly, what mathematical or psychological reasons did the pupil have for arriving at his conclusion? The belief is a symptom of what may be a serious disease or even more than one disease. By substituting numerical examples, or by providing proof, the teacher has dealt with the symptom only and left the disease to reach epidemic proportions.

The following assignment was given to students in a 'methods of teaching mathematics' course. This course was taught in conjunction with Professor S. I. Brown at the State University of N.Y. at Buffalo. (For a more detailed description of this course, see G. Rising: *'Minimum Content For a One Semester Mathematics Methods Course'* in press.)

(i) Justify each of the following classical mathematical mistakes in as many ways as possible, *given* that you believe each is correct.
(A) $(a + b)^2 = a^2 + b^2$
(B) $a/b + c/d = (a + c)/(b + d)$
(C) $\emptyset = \{\emptyset\}$
(ii) Re-examine the justifications in part one and discuss what you believe to be the reason (mathematical or psychological) behind the mistakes.
(iii) Make those mathematical mistakes a positive experience for the pupil(s).

Some of the following justifications are a direct result of the work done on this assignment.

A $(a + b)^2 = a^2 + b^2$
(i) The distributive law of squaring over addition.
(ii) Induction: $(a + b)^1 = (a^1 + b^1) => (a + b)^2 = a^2 + b^2$.
(iii) Consider the Pythagorean theorem, $a^2 + b^2 = c^2$. If $c = a + b$, then $a^2 + b^2 = (a + b)^2$.

(iv) Say the following sentence fast. The sum of the squares equals the square of the sum. The attraction this has is that it *sounds correct.*

(v) Since $(ab)^2 = a^2b^2$ and since multiplication is just a quick form of addition, then . . .

(vi) How many times have we as teachers told our pupils, "Whatever we do to the left side of an equation we must also do to the right side?" Start with the following:

$$(a + b)(a - b) = a^2 - b^2.$$

Now apply the above principle by changing the negative sign in the left side of the equation to a plus, and changing the negative sign on the right side to a plus. Then $(a + b)(a + b) = a^2 + b^2$.

All six of these explanations have two things in common. One is that each has used some principle learned previously and the other is that each has the compelling notion that "in mathematics, the future must look like the past". (Special acknowledgement for this observation goes to Prof. S. I. Brown.)

B $(a/b) + (c/d) = (a + c)/(b + d)$

(i) This looks like multiplication:

$$(a/b)(c/d) = (ac)/(bd).$$

and since multiplication is really a fast way of doing addition . . .

(ii) The "baseball analogy" was first proposed by Dorothy Buerk who assisted, along with others, in the teaching of the methods course. Anyone who follows baseball knows that if a batter has 3 hits out of 5 attempts on Monday and has 2 hits out of 3 attempts on Tuesday, his combined record is 5 hits out of 8 attempts. Therefore it follows that $(3/5) + (2/3) = 5/8$, or in general terms . . .

C $\emptyset = \{\emptyset\}$

(i) The null set is nothing and the set of nothing is nothing, therefore 0 = {0}.

(ii) Notation argument: All other sets use brackets therefore we must use brackets in this case.

(iii) One student claimed that $0 = \{0\} = \{\ \}$, but {0} is best because it leaves *"no doubt"* that we are talking about the empty set. Redundancy stresses the point.

Again, in examples B and C there is that strong need to make the future look like the past.

The multiple reasons given for each mistake suggest that a simple correction by the teacher is insufficient. For example, if a student gives induction as his reason that

$$(a + b)^2 = a^2 + b^2,$$

then a careful examination, or re-examination (whichever the case may be), of the induction principle may be in order. Compare this with the belief that multiplication is a quick form of addition, which many students still have, with good reason. Early in their mathematical career they are introduced to multiplication of integers as an accelerated addition process.

$$2 \times 3 = 2 + 2 + 2 \text{ or}$$
$$3 \times 2 = 3 + 3$$

They are not told that this only holds for a limited domain. When extending the number system they are usually *not* taught that this definition of multiplication needs revision. This can lead to the generalization, as illustrated by the mistake in justification (v), that $x \cdot x = x$ plus itself $x$ number of times, but this principle does not hold when $x$ is not a positive integer. Let $x = \frac{1}{2}$; then by the definition above, $(\frac{1}{2})(\frac{1}{2}) = \frac{1}{2}$ plus itself $\frac{1}{2}$ number of times. Clearly, our definition of multiplication needs some fixing. Negative integers can create a similar difficulty.

Justification (vi) offers yet another disease, the pupils' tendency to generalise rules that are used rather loosely in the classroom. As you can see in this example the rule was applied rather ingeniously.

Examples B and C offer rich ground for determining and perhaps beginning to diagnose mathematical diseases. Perhaps a list of questions will help us begin to re-examine what may be at the root of the mistakes.

1. Where in these justifications is the notion of variable misunderstood?
2. The baseball analogy demands that we take a closer look at so-called everyday occurrences and attempt to see what they may imply mathematically.
3. How confusing is notation?
4. How much do we as teachers underestimate the ability of our students to derive new mathematical principles, e.g. the 'no doubt' principle?
5. Some of the justifications of $0 = \{0\}$ imply that there is a confusion between the meanings of 'is an element of' and 'is a subset of'. How does this misconception relate to other misconceptions students have in beginning set theory?

Let us move away from the 'right versus wrong' conception of a mistake, and from the activity of diagnosing. Mathematical mistakes can be used:

1. to develop new mathematics;
2. to turn a negative experience (making an error) into a positive experience.

Robert Carman in his article *Mathematical Misteaks* has examined in part this type of experience. He limits his work to "mistakes" that have a false justification, but yield a correct answer; e.g.

$$\frac{1\not{6}}{\not{6}4} = \frac{1}{4}.$$

Within this framework he has developed some interesting mathematics. But this development of interesting mathematics can come from any type of mistake, or for that matter a 'forced mistake'.

A 'forced mistake' can best be illustrated with an example. A student teacher, Patricia Burke, taught a unit on Euclidean constructions. After spending time discussing constructions of specific regular polygons inscribed in a circle, she asked her pupils to construct a nine-sided regular polygon inscribed in a circle, *without* telling them that it was impossible. Most of the pupils developed 'solutions'. From there Ms. Burke indicated mathematically that the task was impossible. But that did not stop the class from using those 'wrong' constructions to develop new and legitimate constructions for other types of polygons. New and interesting mathematics was developed and a negative experience turned into a positive one.

What strategies do we have for profiting from mistakes? The answer lies with asking the right questions.

1. Instead of showing pupils that their mistake is wrong, ask when it is right; that is, what questions can we ask to make the solution correct?

   For example, Ms. Burke asked her pupils to develop a question in which their constructions would be possible solutions. They began generating their own questions about geometry and constructions which eventually created an entirely new domain for mathematical exploration. Another example of this type of activity can be seen by exploring mistake B where the pupil adds the numerators and the denominators instead of finding a common denominator. Consider the solution $(a + c)/(b + d)$. In addition to reviewing addition of fractions one can ask, when *can* we add the numerators and denominators together in such a fashion? Pupils can now explore questions such as:

   (i) how many ways are there to find a fraction between two other fractions on a number line?

(ii) is $(a + c)/(b + d)$ *always* between $a/b$ and $c/d$?

(iii) can we talk about betweenness in two or higher dimensions?

2. If this solution were correct, how does this affect the rest of mathematics? And similarly what changes have to be made in our system to accommodate our solution?

For example, let us consider again the error, $(a^2 + b)^2 = a^2 + b^2$. How can we change the system to accommodate this or similar expressions? One consequence is that when two *different* numbers are multiplied together their product is zero. This eliminates the $2ab$ term in $(a + b)^2$. How does this change effect the rest of our system? What happens to $(a + b)^3$, $(a + b)^4$, etc?

3. Are there other *existing* mathematical systems in which our solution holds?

A classic example can be found in the field of geometry. There are many constructions that cannot be done by Euclidean methods and can be done by non-Euclidean methods.

There are many other questions and strategies to be explored; these were just a few. We could do well to re-examine our pupils' mistakes and initiate investigations that will no doubt lead to new pedagogical experiences for ourselves, and positive learning experiences for both teachers and pupils.

## REFERENCE

Carman, R. A. (1971) Mathematical Misteaks. *The Mathematics Teacher, 54,* 109–115.

# 16 Algebraic Explorations of the Error $\dfrac{1\cancel{6}}{\cancel{6}4} = \dfrac{1}{4}$

Raffaella Borasi

Errors can serve as a starting point and as a source of motivation for interesting mathematical explorations. When an obviously incorrect procedure yields a correct result, we may feel puzzled and curious to know how and why this could have happened. Trying to answer these questions can involve us not only in problem solving but in problem-posing activities as well. This experience can provide the opportunity for creativity even in elementary mathematics.

The object of our investigation will be the following well-known "simplification":

$$\frac{1\cancel{6}}{\cancel{6}4} = \frac{1}{4}$$

Why does such an absurd simplification produce the correct result? Is this example the only case for which this kind of simplification works?

We can attempt to answer both questions at one time, by stating the more general problem: For what values of the digits $a$, $b$, and $c$ is

$$\frac{10a + b}{10b + c} = \frac{a}{c}?$$

Or, equivalently, what are the integral solutions between 1 and 9 of the following equation?

$$(1) \quad (10a + b)c - a(10b + c) = 0$$

The values *(a, b, c)* = (1, 6, 4) satisfy this equation, which explains why the result of the simplification turned out to be correct in the specific case presented.

Do other solutions exist? How can we search for them? We do not have a straight-forward algorithm that can be applied to solve equations of this kind, but we can try several approaches.

Given the limited range of our variables (they must be digits), we can create a computer program to check all the possible combinations of *a, b,* and *c* for those that satisfy equation (1). The following BASIC program, for example, will print all the possible solutions in a few seconds:

```
10 FOR A = 1 TO 9
20 FOR B = 1 TO 9
30 FOR C = 1 TO 9
40 IF (10 × A + B) × C = A × (10 × B + C) THEN PRINT A, B, C
50 NEXT C
60 NEXT B
70 NEXT A
```

However, we may not have access to a computer. More important, this "computational" approach does not lend any insight into the nature of the solutions. We are left wondering why some combinations of *a, b,* and *c* "work" and why most others do not. It seems worthwhile, therefore, to try to approach the problem in a more algebraic way.

Just by looking at the example 1∅/∅4, we realize that the simplification will work "trivially" whenever $a = b = c$. We cannot reasonably hope, however, to find other examples by blind trials.

How should we proceed? Frequently, expressing an equation in different but equivalent forms has logical as well as psychological advantages.

We can, for example, try to rewrite equation (1) in different ways to see if anything may be revealed. For example:

$$(2) \quad 10a(b - c) = c(b - a)$$
$$(3) \quad 10ab = c(9a + b)$$
$$(4) \quad 9ac = b(10a - c)$$

Equation (2) may present some advantages, as all *a,* $| b - c |$, *c,* and $| b - a |$ must be less than 10. We can then observe that since 5 divides the first side and 5 is a prime number, either $c = 5$ or $| b - a | = 5$. In our example, we had, in fact, $b - a = 6 - 1 = 5$. We can now see if $c = 5$ in some solutions. With this extra condition, equation (2) becomes

$$10a(b - 5) = 5(b - a)$$

or

$$(5) \quad b = \frac{9a}{2a - 1}.$$

Computing from (5) the values of $b$ corresponding to $a = 1, 2, \ldots, 9$, we *do* find two new solutions besides a trivial one:

$(a, b, c) = (1, 9, 5)$     $\dfrac{1\cancel{9}}{\cancel{9}5} = \dfrac{1}{5}$

$(a, b, c) = (2, 6, 5)$     $\dfrac{2\cancel{6}}{\cancel{6}5} = \dfrac{2}{5}$

We have thus found all the possible solutions with $c = 5$. If other solutions exist, they must derive from $| b - a | = 5$, that is, when either $b = a + 5$ or $a = b + 5$. At first sight checking this case may seem more complicated than checking $c = 5$, but it is actually less so. For $b = a + 5$, equation (2) becomes

$$10a(a + 5 - c) = 5c$$

or

$$(6) \quad c = \frac{2a^2 + 10a}{1 + 2a}.$$

And this time we have only to check for $a = 1, 2, 3, 4$ in (6), as it must be that $b = a + 5 < 10$. We thus find two nontrivial solutions, one of which is our original one:

$(a, b, c) = (1, 6, 4)$     $\dfrac{1\cancel{6}}{\cancel{6}4} = \dfrac{1}{4}$

$(a, b, c) = (4, 8, 9)$     $\dfrac{4\cancel{9}}{\cancel{9}8} = \dfrac{4}{8}$

In the case of $a = b + 5$, equation (2) becomes

$$10(b + 5)(b - c) = -5c$$

or

$$(7) \quad c = \frac{2b^2 + 10b}{9 + 2b}.$$

Checking for $b = 1, 2, 3, 4$ (again it must be that $a = b + 5 < 10$) in (7), we find no other solution.

In conclusion, with only seventeen "checks" (instead of the $9 \times 9 \times 9 = 729$ we might have supposed necessary at the beginning), we have been able to find all the cases in which the "outrageous" simplification would work.

Although we have now solved our original questions, we might still be puzzled by the fact that in all the nontrivial solutions found, $b$ is a multiple of 3. A look at equation (4) can provide some justification for this unexpected result. As 9 divides the first side of the equation, we can deduce that *either* $(10a - c)$ is a multiple of 9 *or* $b$ is a multiple of 3. This does not mean that "$b$ is a multiple of 3" is a necessary condition for a set of solutions of equation (4)—a counterexample is provided by the trivial solution $a = b = c = 1$. However, we can easily prove, by using some divisibility considerations, that the only cases in which $(10a - c)$ is a multiple of 9 is when $a = c$. Since $10a - c = 9a + (a - c)$ and 9 divides $9a$, 9 will divide $(10a - c)$ if and only if it divides $(a - c)$. With the given restrictions on the variables, this means $a - c = 0$. Therefore, all nontrivial solutions must have $b$ as a multiple of 3.

Let us now list briefly some mathematical and educational considerations about this situation that may be of interest to algebra teachers.

The curiosity for finding out when the simplification would work provided a motivation for stating and solving an equation.

We dealt with an equation with more than one unknown and with limitations on the range of the variables. This situation can easily occur in applications, but it does not generally receive enough attention in school. As no "sure" algorithm was available, finding all solutions, or even just some of them, involved some creative problem solving.

The procedure used to solve the original equation clearly pointed out that although logically equivalent, the different ways in which an equation can be written may have specific roles in the search for, and analysis of, solutions. The argument that allowed us to limit considerably the values to check could, in fact, be based only on equation (2). Equation (4), however, helped yield further explanations of the results obtained.

Problem posing is essential in the analysis of this error and can take various forms. In the beginning we had to state the problem mathematically—by formulating the first equation. The absence of an algorithm made it necessary to ask several "unusual" questions, such as, How can we eliminate some values to check? What values are more likely to give solutions? Even when the original problem was solved, we felt the urge to pose a new question: Why does $b$ turn out to be a multiple of 3?

This situation can become a rich source of new problems once we challenge the way we have previously stated the problem (equation 1) or modify some of its elements. For example, we have implicitly assumed that the numbers were written in the usual decimal notation. What if the base of numeration were not ten but another natural number $k$? The problem would then be to find the integral solutions between 1 and $(k - 1)$ of the equation

$$c(ka + b) - a(kb + c) = 0.$$

It may be interesting to discuss the values of $k$ to which we can still apply the argument used in this paper (with proper modifications).

In this article, we have also limited our consideration to two- and one-digit numbers. Can we come up with analogous "simplifications" using more digits? For example, what about $5\cancel{9}4/2\cancel{9}7 = 54/27$? Finding all "three-digit fractions" that can correctly be simplified in this way will now involve a lot more cases. Even if we use a computer, we will face the real challenge in writing an efficient program and eliminating a priori as many trivial solutions as possible (you can expect hundreds of solutions in this case!). What are other possible simplifications that can occur with "three-digit fractions"? What is the percentage of "correct" versus "wrong" results of each simplification? Does any pattern occur in the solutions?

This problem can provide concrete material and the stimulus for a discussion about the difference between necessary and sufficient conditions for solutions and about the values and limitations of heuristic procedures versus algorithms in solving equations. It can also provide further reflection on the use of computers in mathematics, in comparison to more "classical" mathematical activities.

## REFERENCES

Brown, S. I., & Walter, M. I. (1983). *The art of problem posing*. Philadelphia: The Franklin Institute Press.

Carman, R. A. (1971). Mathematical Misteaks. *Mathematics Teacher, 54*, 109–115.

Meyerson, L. N. (1976). Mathematical Mistakes. *Mathematics Teaching, 76*, 36–40.

# Tinkering With What Has Been Taken For Granted Editors' Comments

Textbooks seem to take for granted that students from the earliest age on have to be supplied with the problems.

In **Problem Stories: A New Twist on Problem Posing** Bush and Fiala encourage fifth grade students and pre-service elementary school teachers to write their own problem stories. They discussed with their students ways of making problems easier or harder—a nice additional problem posing strategy—one we did not mention in **The Art of Problem Posing.** They end the article with useful advice for preparing students to write their own problems. This article is interesting on several counts. First, problem solving is often equated with solving word problems—and they are often of a very stereotyped kind. By encouraging and helping students to pose their own story problems students have a good chance to break out of this mold and also to learn what information needs to be given in order to solve a problem. In **The Art of Problem Posing,** we suggest a HANDY LIST of (helpful) questions and begin a sample list. Bush and Fiala pose a number of questions that students could very profitably add to their own list of handy questions. Among them are "Could we solve this problem with less information? What other information is needed to solve the problem? Is just the right information given to solve this problem?"

Very often we see problems in student texts and take them for granted. They are to be solved and left at that. In the following pieces Brown; Moses, et al; and Walter create and illustrate ways of turning even routine exercises into problem posing activities.

In **How to Create Problems,** Brown gives several examples of how an innocent looking ordinary exercise such as "Compute the total value of 3 nickels and 4 dimes" can lead to a host of new problems at a variety of levels. He suggests *focusing backwards* — start with the answer 55 cents — what questions might you raise? And he suggests *focusing on the interrelationship* between some of the aspects of the original exercise in order to generate new ideas. Another way of interpreting *focusing backwards* is to think of *a reversal of view point.*

Moses et. al. in **Beyond Problem Solving: Problem Posing** also start with a coin problem and then they elaborate upon other problem posing strategies. They suggest the following strategies for problem posing: Consider what is known, unknown, and consider any restrictions. Then ask yourself what if the known and unknown were reversed and what if the restrictions were changed? They give several practical suggestions for easing students into the problem posing mode. They also discuss a classroom example in which the students examined and posed problems starting with a number lattice. Lattice work is rich indeed especially if one asks what if the lattice is not rectangular.

Kissane, in **Mathematical Investigation: Description, Rationale, and Example,** starts by discussing what he means by mathematical investigation and goes on to give a rationale for "reserving space in the curriculum for mathematical investigation."

Kissane then takes an innocent looking open situation as an example: "Investigate the numbers adding to thirteen." From it he is led to the question: "How many partitions of thirteen are possible?" He goes on to show how the What-If-Not technique can lead to further new questions. He shows how he arrived quite naturally at the question "Which numbers can be expressed as the sums of consecutive integers?" Kissane presents a beautiful example here of how an interesting problem can arise from a simple starting situation. He enriches this latter problem by interchanging the hypothesis and the conclusion — a useful problem posing strategy. You might want to think about how this process compares to the ones used by Brown and Moses et al. Then, using the What-If-Not technique again, he shows how he is led to the problem "Which partitions give the greatest product?". Investigating this question leads him to several conjectures. He points out how the very resolution of a problem can lead to new questions.

Kissane then shows how, by tinkering with what we often take for granted, namely the domain—in this case the integers—we can get further insight into a problem. He also illustrates the fact that problems are not always clearly stated and thus need to be refined. The result he obtains after discussing in detail how he gets to his final conjecture is so surprising that we will not divulge it here.

In **Curriculum Topics Through Problem Posing**, Walter does not take for granted that all arithmetic exercises have to be *given* to students and shows how routine exercises can be turned into problems by noticing any characteristics of an arithmetic exercise and asking students to make up other problems with answers that have this characteristic. For example: Given an addition problem involving two three digit numbers whose sum happens to be 876, can you make up other addition problems whose answers consist of three consecutive numbers. In this way practice can be combined with problem solving and with problem posing.

In **Is the graph of y = kx Straight**? Friedlander and Dreyfus present us with a collection of fascinating locus problems—problems that are accessible to youngsters at a much earlier age than is usually the case. In order to make the problems more accessible, they supply their students with a number of pre-set coordinate systems rather than accept the standard Cartesian coordinate system for all locus problems. Given a point P(x,y), if x and y represent the *distances* from P to two fixed perpendicular lines then the graph of the equation y = kx is two straight lines passing through the origin. If, however, x and y represent the distances from P to a non-Cartesian system (say a fixed point and a fixed line rather than two perpendicular lines) then the graph of y = kx is different; if k = 1, for example, the graph is a parabola.

By clever use of What-If-Not ideas on modified coordinate systems, the students becomes sensitive not only to the effect of variations on the Cartesian coordinate system, but they acquire new appreciation for the standard coordinate system they have accepted all along. Friedlander and Dreyfus demonstrate powerfully how it is that one understands a concept more deeply when one sees the effect of changing it. Their article, which integrates geometric and algebraic thinking, is a good transition piece from chapter II to chapter III.

# 17 Problem Stories: A New Twist on Problem Posing

William S. Bush
Ann Fiala

Problem solving has become the focus of the '80s. The *Arithmetic Teacher* and the *Mathematics Teacher* are full of articles on problem solving; conferences for mathematics teachers overflow with sessions on problem solving; and more and more teachers of mathematics are jumping on the problem-solving bandwagon. If you are one of these teachers, this article should interest you.

We have been involved with problem solving for several years — one with classes of fifth graders and the other with preservice teachers at the university level. To cap off a year of problem solving and to challenge our students further, we contrived the activity of writing problem stories — a new twist on problem posing. In writing problem stories, students compose original stories that incorporate original nonroutine problems. The problems must contribute logically and naturally to the plot of the story. Successful completion of the task requires excellent writing skills, creative thinking, and adept problem-solving ability.

As a classroom activity, problem posing has several inherent strengths. Brown and Walter (1983) suggest that problem posing greatly enhances the understanding of problem solving, develops a better understanding and different perspective of phenomena around us, helps alleviate fear of mathematics, and dispels the "right way" syndrome in mathematics. Wirtz and Kahn (1982) add that having students make up applications helps them bridge the gap between concrete situations and mathematical abstractions, helps them learn to generalize, and makes mathematics more meaningful to them. Writing problem stories has the additional advantages of developing creative writing skills and integrating mathematics with other subject areas (see Jason LaGrega's problem story for an example in history).

Writing problem stories is an appropriate activity for students of all ages.

To illustrate its flexibility and diversity, we share two contexts in which we assigned problem stories, two sample stories, and suggestions for assigning problem stories.

## FIFTH-GRADE CLASS

For the foregoing three months, problem-solving activities had challenged my fifth-grade gifted-and-talented students to analyze situations, synthesize data, propose strategies, and evaluate the feasibility of solutions. Initially I gave the students two or three nonroutine problems to be solved individually or in pairs during class. At the end of these sessions, we shared solutions and compared strategies. After having the students solve various types of problems for the next few weeks, I gave them a list of sixteen problem-solving strategies taken from *Problem Solving: A Basic Mathematics Goal—Book 1* (Ohio Department of Education 1980). Next I gave students problems from that text that required particular strategies for solving. Through solving these problems, students soon learned that some solutions required a combination of strategies. I then returned to sets of varied problems for which students would have to select strategies.

To prepare my students for writing problem stories, I often asked them to generate new problems similar to previously solved problems at the end of our problem-solving sessions. We also discussed ways in which problems could be made easier or more difficult. Several days prior to the assignment, we reviewed the literary elements (setting, characters, plot, theme) of a short story. We brainstormed the many possibilities for each element and discussed the types of stories that might be conducive to incorporating problems. Adventure stories were used to illustrate ways in which plots can depend on the solution of a problem.

Following these discussions, the students were asked to write a problem story. They were required to include at least two original nonroutine problems of different types and to submit the solutions and the strategies for their problems. They could work individually or in pairs. Jason's story, "The Battle at Richmond," represents one of the more creative stories.

# THE BATTLE AT RICHMOND

The Civil War was near its end and everyone was dreary and tired. Both the Confederates and the Yankees knew how important it would be to gain control of Richmond. Lee and Grant both wanted to have it because the supply trains for both armies ran through there.

Lee was marching 15 000 men to Richmond and Grant was marching 25 000 men there. Grant and Lee didn't have precious time to waste.

Lee had the choice to go through 3 marshes and 1 forest to get there. It would take 2 days to get through the forest and 2 more days to get through the marshes. Lee could choose 4 forests instead of the 3 marshes and 1 forest. 3 forests were each 1/4 the size of the 2-day forests, and the last forest was twice the size of the 2-day forest. Lee decided to take the 4 forests instead because the marshes were too dangerous.

Grant had a choice to go through 2 forests. One would take 1/3 of Lee's 2-day forest and the other would take 2 times more time than Lee's 2-day forest. He could also choose 4 swamps. It would take 1/4 of a day to go through one swamp and another one would take 1/2 day. It would take 5 more days to go through the other 2 swamps. Grant decided to go through the forests. How long would each trip take, which one is shorter for each, and who would reach Richmond first?

From solving the problems you now know that the Yankees reached Richmond before the Confederates. When the Confederates arrived, the battle began. The Yankees had an advantage because they were ready and waiting. When the Confederates saw what was happening, they hid behind trees and started digging trenches to protect themselves. Lee had to estimate about how long it would take to dig a trench that was 5 feet deep and 70 feet long. It took 5 hours to dig a trench 5 feet long and 5 feet deep. How long would it take to dig the trench?

The armies were exhausted after 6 days of fighting. The Confederates lost 5000 men and 1113 were injured. The Yankees only lost 2000 men and 300 were injured.

The Confederates were surrounded because the Yankees sent 10 000 men behind them. General Lee surrendered after 3 weeks of fighting. He surrendered because he only had 4327 men left, and that ended the Civil War and brought the United States back together.

*Answers:* Lee's first choice—4 days; Lee's second choice—5 days, 12 hours; Grant's first choice—4 days, 16 hours; Grant's second choice—5 days, 18 hours; ditch could be dug in 2 days, 22 hours. *(Jason LaGrega, Fifth Grade, DeLeon Elementary School, Victoria, Texas)*

## PRESERVICE ELEMENTARY SCHOOL TEACHERS

Twenty preservice elementary school teachers were enrolled in a course on mathematical problem solving. The purpose of the course was to develop the problem-solving skills of the preservice teachers through solving nonroutine mathematics problems. The course textbook (Averbach and Chein 1980) focused on problems involving logic, algebraic equations, Diophantine equations, number theory, cryptarithms, networks, and graph theory. I assigned selected problems from the textbook, as well as problems taken from other sources, throughout the semester. The majority of class time was devoted to the sharing of solutions to problems at the blackboard. The few class presentations I made focused primarily on the mathematics prerequisites needed to solve the problems.

The problem story was an optional assignment made at the beginning of the semester and due at the end. Preservice teachers who chose this option were required to write a story that incorporated five original nonroutine problems, each of a different type. Since the stories were quite lengthy, excerpts have been taken from one of the better stories, entitled "The Untypically Busy Day of the Christensen Family," by Sue Anne Shahan.

## THE UNTYPICALLY BUSY DAY OF THE CHRISTENSEN FAMILY

One morning, Mrs. Christensen looked at her calendar and realized that it was time for her children to have checkups. Mrs. Christensen knows she must carefully plan trips to town. Many errands have to be run, and Mrs. Christensen tries to use her time in town very efficiently. Planning this particular trip to town would take more time because this time all four children would be going with her.

Because the dentist and the orthodontist were husband and wife, they tried to keep the same hours. Both were opened on Monday, Tuesday, and Friday, the latter also being open a half a day on Thursday. The only weekday the town doctor was closed was Wednesday. The optometrist was closed on Wednesday and only takes care of emergencies on Monday. Of course, none of the offices were opened on Sunday.

It took Mrs. Christensen a few days and several telephone calls, but she was eventually successful in scheduling appointments for all four children on the same day. On the evening before the big trip to Fort Morgan, Mrs. Christensen reminded the children that they would all need to be ready to leave the house at 7:00 A.M. in order to get to the first scheduled appointment, which was at 9:30 A.M. She promised them that if they were all ready to leave at 7:00 A.M., she would treat them to ice cream after lunch.

The idea of eating ice cream seemed to give everyone the motivation to be ready on time. To Mrs. Christensen's surprise, they were in the car and ready to leave at exactly 7:00 A.M. When they arrived in Fort Morgan, it was six minutes before the first appointment. Oh, how pleased Mrs. Christensen was that the morning had gone so smoothly.

At noon, Mrs. Christensen dropped the bunch off at the malt shop for lunch. The children were to eat their lunches and be ready in forty-five minutes to finish the errands. While munching on their lunches, the following conversation took place:

Brian: My appointment could have been yesterday. We all could have come tomorrow.

Bruce: There are only two days that all offices are closed. Only two of us could have come tomorrow.

Dana: There is only one weekday my appointment could not have been. Brian could not have come yesterday.

Janis: We all could have come yesterday. I could have come both yesterday and tomorrow.

Now, two of the Christensen children always spoke the truth and two of them always lied. If each child had an appointment with a different doctor, which two told the truth, which two were liars, and who had what appointment? . . .

Finally, all the errands had been run and all the appointments had been met. Mrs. Christensen rounded up her crew and headed for home at 3:00 P.M. All of a sudden a severe thunderstorm was upon them. Visibility was extremely limited, causing Mrs. Christensen to cut her regular speed in half. When the trip home was halfway completed, the storm lifted and Mrs. Christensen was able to resume her normal speed. At what time would she get home?

*Answers:* Brian and Janis are lying; Bruce and Dana are telling the truth; Brian — dentist, Bruce — optometrist, Dana — doctor, Janis — orthodontist; arrived home at 6:36 P.M. *(Sue Anne Shahan, senior majoring in elementary education, University of Houston — Victoria, Victoria, Texas)*

## SUGGESTIONS

In assigning problem stories to our classes, we have learned some helpful approaches. We would like to offer several suggestions for preparing students for the task of writing problem stories.

*1. Have students solve many types of nonroutine problems.* Students need to know what a problem is, how to go about solving problems, and what feasible solutions look like. This can only be accomplished by actively involving your students in problem solving.

*2. Once problems have been solved, encourage students to create similar problems.* This suggestion stems from Polya's (1971) looking-back stage in teaching problem solving. Ask students how previously solved problems might be made easier or more difficult or what information in a problem might be changed.

*3. Have students create problems for existing stories.* This activity permits students to focus only on the problem-posing aspect of problem stories. Adventure stories function well in this capacity.

*4. Give students ample opportunity to write short stories.* This activity permits students to focus only on the writing aspect of problem stories. Devote time to analyzing and reviewing short stories and their components.

*5. Have students work together in small groups.* Working in groups generally reduces anxiety and frustration and encourages cooperation. Getting students to brainstorm a story or the problems can make the task easier for them. You might have students take turns creating stories, then posing problems to fit them or posing problems, then creating appropriate stories.

*6. Encourage students to create both revised problems and new problems.* Making simple revisions of previously solved problems can be a challenging task for many students. However, some students need the challenge of creating new problems. The more capable students should be encouraged to create original problems rather than to restrict themselves to revised problems.

*7. Allow students to rewrite their stories and revise their problems.* A better product will result if students are allowed to edit their work. You might also encourage students to exchange stories and edit each other's work.

*8. Encourage students to solve other students' problems.* Problem stories can provide your class with an endless supply of problems at an appropriate level of difficulty. This activity also brings us full cycle—back to problem solving.

*9. Collect problem stories in a class anthology.* Have the students select their best efforts and place them in a class anthology to be presented to parents or other classes.

Writing problem stories has been an extremely successful activity for both our classes. Our students have enjoyed writing and posing problems, and we have delighted in reading the stories and solving the problems. These stories have shown us that there are no bounds to students' creativity, imagination, and ability. We wish you success in having your students write problem stories.

## REFERENCES

Averbach, B., & Chein, O. (1980). *Mathematics: Problem solving through recreational mathematics.* San Francisco: W. H. Freeman & Co.

Brown, S. I., & Walter, M. I. (1983). *The art of problem posing.* Philadelphia: The Franklin Institute Press.

Ohio Department of Education (1980). *Problem solving: A basic mathematics goal—Book 1.* Palo Alto, CA: Dale Seymour Publications.

Polya, G. (1971). *How to solve it.* Princeton: Princeton University Press.

Wirtz, R. W., & Kahn, E. (1982). Another look at applications in elementary school mathematics. *Arithmetic Teacher, 30*(1), 21-25.

# 18 How to Create Problems

Stephen I. Brown

We have distinguished between exercises and problems. In this section, we hope to show how *you* can become a source of good problems by creating them on your own. Indeed, we hope that *you* will find as much intellectual excitement in creating problems for your students as they will find in solving them. Perhaps you will eventually want to teach your students some of these strategies for creating problems.

We begin by taking exercises at different levels of sophistication in the elementary grades and turning them into problems.

## CASE 1: "MONEY MAKES THE WORLD GO AROUND"

*Exercise:*  Compute the total value of 3 nickels and 4 dimes.
*Solution:*  $(3 \times 5) + (4 \times 10) = 55$ Answer: 55 cents

How might you transform the exercise into a problem? One good way would be to *focus backwards*. For many students, finding the *total* value of a handful of coins is a simple exercise. Though they may make many mistakes in arithmetic, the task is conceptually very simple. Focus *now* on the 55 cents instead of the coins, and ask some questions. Here are a number of questions you might ask:

*Question 1:*  Could 55 cents be made up only of nickels?
*Question 2:*  Could 55 cents be made up of dimes?

So far, so simple, though for *some* youngsters, questions 1 and 2 would constitute problems. Can you find ways of creating a more challenging question involving 55 cents? So far we have looked at nickels and dimes separately. How would you ask a question that might combine them?

*Question 3:*   What other combinations of nickels and dimes can you come up with to get 55 cents?

Now that we have combined nickels and dimes, what other questions might you ask that could place demands on their *interrelationship*? Below is one example.

*Question 4:*   What is the largest number of dimes you could have if you had nickels and dimes that totaled 55 cents?

So far, we have used only nickels and dimes to yield 55 cents. What else might we do to make this a more challenging activity?

*Question 5:*   Suppose you have nickels, dimes, and pennies that add up to 55 cents. Could you have all pennies? All nickels? All dimes?

How can we interrelate the three kinds of coins and come up with additional questions? As in the case of nickels and dimes, you realize that only certain combinations work. Realizing this, you might ask these additional questions:

*Question 6:*   What is the greatest *number* of coins that can make 55 cents?
*Question 7:*   What is the least *number* of coins that can make 55 cents?
*Question 8:*   Is there any number of coins between the smallest and largest number that will *not* make 55 cents?

There are, of course, many more challenging questions that you could devise with Case 1 as a starting point. It is possible to begin with something relatively routine and work up to challenging questions. Strategies for solving this problem, and other types of interesting questions that could come out of it, will be considered in section III of this monograph.

Here is another exercise that we can transform into interesting problems.

## CASE 2: "CHICKENS AND COWS"

*Exercise:*   Mr. Brown has three cows and eight chickens. How many legs are there all together?

At this point you should try to compose a few good problems using this basic barnyard setting. Look back at Case 1 for similarities.

## CASE 3: SOME WORK WITH RODS

Our work with algorithms aims to provide us with efficient ways to add, multiply, subtract, and divide numbers. In the lower grades, children frequently make use of concrete material. For example, youngsters persuade themselves that $8 + 2 = 10$ by showing that a rod of length 8 joined to one of length 2 has the same length as a rod of length 10.

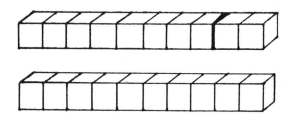

Suppose you employ the strategy suggested in Case 1 and Case 2, however. *Focus backwards.* Instead of looking first at the *components* of a sum, focus on the sum. Consider the following question: Which pairs will join to match the 10 rod? Clearly, the 2 and 8 rods will work. What else will work? We might be interested not only in those pairs that work but in the *numbers* of pairs. The following question seems reasonable: How many different pairs of rods can we find to match the 10 rod?

Recall that in Case 1 we moved from nickels and dimes to include other denominations. How would that suggest a reasonable question, still maintaining the 10 rod as a focus? There are many possibilities for you to explore.

## CASE 4: ODD OR EVEN?

We have all memorized multiplication tables as youngsters. The "two times" table for example, begins, "2, 4, 6, 8, 10, 12, 14, 16 . . ."

A *reversal* of the viewpoint enables us to come up with many questions — some of which appear in every text and others of which await *your* insight and creativity to be asked. Let us start with an easy one:

Given any number, such as 57832, does it belong to the "two times" table?

You can come up with *many* interesting additional questions.

Suppose someone has written a five-digit number that allegedly belongs to the "two times" table. Part of it was erased. How much of the number would have to be left intact for me to agree that the number belongs to that table?

We know that *not* all numbers are even (belonging to the "two times" table). That should suggest many questions that might be challenging for some people who are first beginning to think about even numbers. For example:

If you add two even numbers, what do you get?

If you add two odd numbers, what do you get?

Ask four more questions based upon even and odd numbers in combination with some operations of arithmetic. The best questions are those whose answers you do not know.

## SUMMARY

Though we have done a number of things in this section in order to suggest how problems can be *created,* we have for the most part focused on one overall strategy—that of reversing our focus. Instead of taking what was "given" for the purpose of reaching a conclusion, we have taken the *conclusion* and wondered about what might have gotten us there. It was easy to figure out that if we are given a certain number of coins of different denominations, then we would have a certain amount of money. The reverse focus, however, led to a collection of interesting problems. If we are given a certain amount of money, what kinds of coins might we have? Do you see how the concept of reversing what we are given and what we can conclude leads to a whole new set of interesting problems in each of the four cases?

# 19 Beyond Problem Solving: Problem Posing[1]

Barbara Moses
Elizabeth Bjork
E. Paul Goldenberg

How many coins does it take to make 45¢, using just nickels?"
Nine . . .
. . . And then we are finished.

## WHAT IS PROBLEM POSING?

As the problem above stands, it is fully specified: a simple question with a
unique answer. But behind each such problem, no matter how limited it
may seem, lurks a world of other potentially interesting problems. One
strategy for getting to these problems is to ask what *kind* of information the
problem gives us, what *kind* of information is unknown (and wanted), and
what *kinds* of restrictions are placed on the answer. An analysis of the coin
problem is given in the chart shown in Fig. 19.1.

|  | Kind of Knowledge | Details |
|---|---|---|
| **Known** | The sum of money | 45¢ |
| **Unknown** | The number of coins | |
| **Restriction 1** | A *particular* coin must be used at least once. | 5¢ |
| **Restriction 2** | All coins must be of the same type. | |

FIG. 19.1.

Many new problems can be generated from our original problem by changing the knowns or unknowns or the restrictions.

- What if we reversed the known and the unknown? That is, what if the *sum* were unknown and the *number* known? A new problem is born: What sum of money can be made using exactly nine nickels? And the problem is still fully specified.
- What if restriction 1 were dropped and the problem did not specify that nickels be the coin used? A new problem is born: How many coins does it take to make 45¢, using just one type of coin? The answer is not unique; it is either 9 or 45, depending on one's choice of coin. If we also reverse known and unknown in the original problem, we have yet another problem: What sum(s) of money can be made using exactly nine coins, all of the same type?
- What if restriction 2 were dropped? A new problem is born: How many coins does it take to make 45¢ using at least one nickel? Again, the answer is not unique; there are seven different ways of making 45¢ using exactly one nickel, and lots more ways using more nickels.
- The implied domain is *all* standard American coins: 1¢, 5¢, 10¢, 25¢, and so on. What if we changed domains and used a different set of coins?
  *What if we didn't have pennies, but otherwise used American coinage? What sums of money could we no longer construct?
  *What if we didn't have pennies, but did have a 3¢ coin (and other American coinage)? What sums of money could we no longer construct? (We would no longer have exact change for 1¢, 2¢, 4¢, and 7¢.) Even though we could not buy certain objects with exact change, perhaps some of them could be bought if we paid more and collected the right change (e.g., paid 10¢ and received 3¢ change). Which ones? (All of them)

## HOW DO WE POSE PROBLEMS?

### Identifying and Changing Constraints

---

**Principle 1:** Have students learn to focus their attention on *known, unknown,* and *restrictions*. Then consider the following question: What if different things were known and unknown? What if the restrictions were changed?

---

The most mundane of problems can become richer if instead of just asking, "How do I solve it?" we first ask, "What's this problem all about?" Principle 1 organizes a child's exploration of what a problem is about and suggests a way to create a new problem.

Brown and Walter (1970, 1983) maintain that this approach is not only learnable but teachable. Students can be taught to expect that mathematical statements and problems, like the one that we explored at the beginning of this paper, include something *known,* something *unknown,* and (often) some *restrictions.* Sometimes these restrictions are subtle: by thinking about stamps instead of coins, we might notice that there are more denominations than 1¢, 5¢, 10¢, and so on. Removing a restriction shows how a problem is part of a larger class of related problems rather than an isolated exercise. Initially, the teacher bears most of the responsibility for helping students see the classes from which each problem instance is drawn. In time, students do more and more on their own, building a repertoire of associations for creative problem posing and effective problem *solving.*

## Looking at Familiar Things in Strange Ways

> **Principle 2:** Begin in comfortable mathematical territory.

By starting in a context that is sufficiently familiar, even very young children — with some encouragement and modeling by the teacher — can list attributes and change the constraints of a problem. This kind of thinking should be begun in the earliest grades to encourage a problem-posing attitude.

For young children, manipulatives often help to make a mathematical territory feel familiar. In our first example, we used coins to give a homey feel to what might otherwise have been a very abstract and arbitrary set of restrictions about addition problems. Brown and Walter (1983) show how even a common manipulative like a geoboard can be used for attribute finding. As they are, geoboards have square borders, pegs at lattice points on a square lattice, and finite size. What if one or more of these characteristics were changed? And what new questions does a change in characteristics lead one to ask?

For example, what if the geoboard's outline were an oblong or a circle instead of a square? We might then ask what such a change of shape does to the number of pegs on the board. For example, how many fewer pegs are on a diameter-2 *circular* segment of a geoboard than on the entire 2 × 2 *square* geoboard? Fig. 19.2 shows circular segments of 2 × 2 through 5 × 5 geoboards. On the smallest of these geoboards, 4 pegs lie outside the circle. The others exclude 12, 12, and 20 pegs, respectively.

Varying the attributes and asking new questions leads to a host of discoveries.

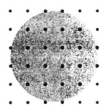

FIG. 19.2.

## Using Ambiguity: Necessary but Insufficient Conditions

> **Principle 3:** Encourage students to use ambiguity to create new questions and problems.

Sometimes children find it easier to express what they *want to know* about a mystery object or an ambiguous statement than to state what they *do know* about something that is completely revealed to them. Hold up a box with some coins in it, tell the students it contains coins, jingle the coins, and encourage students to ask questions about what's in the box. Record all questions. If necessary, stimulate new trains of thought by suggesting questions like "Are they all U.S. coins?"

A common teaching technique aimed at clear communication uses this idea in reverse. The teacher asks students for a definition of, say, a triangle and deliberately misinterprets an incomplete student response such as "It has three lines" by producing a drawing like that shown in Fig. 19.3. This is problem posing par excellence—keeping one feature (the student's specification) fixed while relaxing the other constraints—and so it is a perfect opportunity to model that problem-posing technique. Each incomplete student specification, rather than being treated as wrong, is seen as one essential feature—perhaps insufficient but absolutely necessary. As students get the idea, the teacher may do less of the problem posing and turn that task back to the students: How can we draw a figure that conforms with the feature just mentioned but is not a triangle? Or in the language of "what if," What if it had three lines but was not a triangle?

The goal should be not just to "catch" the ambiguity as if it is merely an error but to *use* it productively. This approach not only de-emphasizes the inadequacy of one student's incomplete specification but has the added advantage of focusing all students' attention directly on the unspecified attributes, helping them learn to recognize those attributes. The amusement

FIG. 19.3.  $J \mid \Gamma$

value of the perverse drawings and the puzzle are maintained, as well as the goal of helping students use more precise communication.

Ambiguity leaves room for curiosity, imagination, and generating one's own ideas. Ambiguity is also inevitable. Where it arises, we should make it useful rather than a failure.

## Making the Domain Explicit

> **Principle 4:** Teach the idea of domain from the earliest grades, encouraging children to "play the same (mathematical) game with a different set of pieces."

Games may help students look at constraints creatively. In some kinds of games (especially ones like chess, nim, go, and 21, which we see as very mathematical), there are playing pieces (pawns, counters, cards, etc.), rules by which these pieces may be manipulated (moving, taking, counting, etc.), and goals to achieve (a particular configuration of pieces, a specified sum, the most or least, being last to move, etc.). By *domain* we mean the mathematical "objects" — they may be specific numbers, geometric shapes, functions, or other mathematical abstractions — that we decide to include in our game.

There are always opportunities to explore different domains that remain within the child's mathematical competence, and from the earliest grades students should be taught to "try the same game with a different set of pieces." Any mathematical task that a child can perform in one domain (e.g., counting numbers) can be explored within a subset of that domain (e.g., evens, or the set $\{1, 4, 7, 10, 13, \ldots\}$). Often, the question "What numbers am I allowed to use?" is sufficient to call the idea of domain to mind. For example, in the problem *Name two numbers whose product is 12,* do we allow all numbers, only integers, only evens, only odds? (In the case of "only odds," of course, the problem has no answers.)

## HOW CAN WE FOSTER CREATIVE PROBLEM POSING?

The teacher is the essential ingredient. It is the teacher who sets the context by helping students learn how to open up one problem to reveal the others it suggests. And it is also the teacher who establishes a classroom climate conducive to spontaneous and productive inquiry in several ways: modeling the process personally by wondering openly *with* the students, fostering the free exchange of ideas and actively encouraging collaboration among students, honoring students' spontaneous what-ifs and conjectures, and being as interested in *how* students thought about a problem as in *what* they came up with.

The following two strategies can foster problem posing among students.

## Use Problems in the Textbook as a Basis for Problem Posing

Using techniques such as those outlined above, the teacher can regularly pick a problem that can be enriched through problem posing. One might begin with the best problems in the text, select problems from supplementary sources (e.g., Greenes et al. [1980], Lane County [1981], and Ohio Department of Education [1982]), or deliberately pick a not-so-interesting problem to explore and subsequently think out loud with the students.

## Avoid Questions That Have Unique Answers

Problems that have more than one solution tend to foster a problem-posing mind-set because they are not as limited as one-answer problems. For example, the questions

"How many coins does it take to make 45¢, using *at least one* nickel?" or, more broadly,

"How many ways can you make exactly 45¢ with U.S. coins?" provide a richer context for problem posing than the following question:

"How many coins does it take to make 45¢, using *just* nickels?"

## Classroom Climate

Although most students are not used to being problem posers, they can learn to be if their teachers model the process for them and create an environment in which they feel free to pose their own problems. We offer the following suggestions.

- *Let students choose what problems they try to solve.* Good problem posers will raise more problems than time permits them to solve. Also, some problems will generate more interest and curiosity than others, and so individual taste and choice will play a part in who follows up which problem and to what extent. Students may also raise problems that they *cannot* solve. Students feel free to pose problems when they do not fear being embarrassed if they invent one that is too hard for them. Some of the most important discoveries in the history of mathematics were inspired by problems that still remain unsolved.

- *Place no time pressure on solving a problem*. Good problems take time to explore and yield many new interesting ideas. For both reasons, good problem posing requires lots of exploration time. Student problems that remain unsolved for whatever reason can resurface from time to time, stimulating curiosity and further thought, and should be seen as challenges, not failures.
- *Brainstorm with the students; encourage communication and collaboration*. Students feel most free to risk posing new problems when they and the teacher work *together* as collaborating partners in problem posing. Encourage the productive exchange of ideas among groups of children and remain flexible about the direction of the lesson and the stream of student inquiry and initiation. Mathematical content remains essential, but when students and teacher focus on the process of solving problems and the challenge of posing new ones, the emphasis shifts from *acquiring* facts to *using* them.

Collaboration among students has many benefits. For one thing, it provides a context in which students must develop ways of communicating about mathematics and build a vocabulary for their mathematical ideas. The *Curriculum and Evaluation Standards for School Mathematics* (NCTM 1989, p. 6) emphasizes that "as students communicate their ideas, they learn to clarify, refine, and consolidate their thinking." Collaboration and communication are particularly important for problem posing, for as students share their different ideas, questions, and perspectives, they fertilize the development of new ideas, questions, and perspectives.

Record and honor spontaneous what-ifs and conjectures. Naming an idea after its inventor—Rachel's method, the Jonathan conjecture—can be a great way of honoring the inventor and encouraging others.

## Using Technology to Promote Problem Posing

Computers and calculators free students and teachers alike from tedious computational tasks, from repetitive numerical and geometric manipulations, and from overwhelming memory jobs. Because the computer calculates and recalculates so quickly, students can easily ask and explore the what-if questions with relatively little tedious computation. In this way, appropriate use of technology can foster and enhance problem posing by students.

In one fourth-grade classroom, the students were asked to find an average score for their recent mathematics test by using a piece of graphing software. Students were able to enter the test data quickly, display bar graphs, and compute the mean score of the class. When a new child joined the class, the students were able to add the new student's score and see how it affected the bar graph and the mean score for the class. They could also ask many what-if questions; for example: "What if two students dropped out of the class?" "What if the teacher had made a five-point error in each student's score?" "How many scores need to be adjusted to have an average of 90?"

The computer allows students and teachers to explore domains that were previously unwieldy or unavailable in a pencil-and-paper world. Data can easily be collected, and the computer will display them in many forms, allowing the student and teacher to make predictions and ask new questions about the effects of changing any one of the conditions. New problems are easily generated because the tool displays graphically that which the students are exploring.

Skilled communication contributes significantly to mathematical learning. Students need to be encouraged to speak and write mathematics and learn how to sense when words are not as effective as pictures, diagrams, symbols, and graphs. The computer seems naturally to encourage collaborative experimentation and the sharing of results of problem posing and solving. When it links graphical and symbolic representations, it also helps students see when each representation is appropriate and then learn to choose among them.

## CLASSROOM EXAMPLE: A LATTICE SETTING

Because of our strong belief that problem posing is central to the learning of mathematics, we have begun writing K-6 curricular materials in which problem posing pervades all the activities. This section describes one area of the *Reckoning with Mathematics* curriculum and the way problem posing arose by (1) using interesting problems, (2) having the teacher model good problem-posing behavior, and (3) letting the students follow their own interests. Notice how the technology enhanced the problem-posing process.

We showed a class of fourth graders a ten-lattice (Page 1965, Goldenberg 1970) as a way of looking at number patterns. Shown an incomplete matrix of ten columns (Fig. 19.4), students were asked to fill in the missing numbers. What number should be next to 32? What number should be above 32? Below 99?

The teacher explained that although the lattice drawn on the board ended at 99, the lattice actually continued. For how long? "Until you die!" asserted one student. The chalkboard is a static medium, but using a piece of prototype software designed by Education Development Center for this module, the teacher and students were able to generate any lattice and scroll it up and down to show how it continued "forever" in both directions.

Because of the rich environment of the lattice and the familiarity of the

| 90 | 91 | 92 |    |    |    |    |    |    | 99 |
|----|----|----|----|----|----|----|----|----|----|
| •  |    |    |    |    |    |    |    |    |    |
| •  |    |    |    |    |    |    |    |    |    |
| •  |    |    |    |    |    |    |    |    |    |
| •  |    |    |    |    |    |    |    |    |    |
| •  |    |    |    |    |    |    |    |    |    |
| 30 | 31 | 32 |    |    |    |    |    |    | 39 |
| 20 | 21 | 22 | 23 | 24 | 25 | 26 | 27 | 28 | 29 |
| 10 | 11 | 12 | 13 | 14 | 15 | 16 | 17 | 18 | 19 |
| 0  | 1  | 2  | 3  | 4  | 5  | 6  | 7  | 8  | 9  |

FIG. 19.4.

domain, the teacher's role at this point was to pose a few open-ended questions. "What patterns do you see? How many rows up would you have to go to reach numbers in the thousands? What could we put below zero?" The question "How would you describe the lattice to someone who had never seen one?" is an attribute-finding question from which the children could begin to pose problems. At the teacher's suggestion that other lattices could be made, the entire class got into the act. "What if we started at 2? Could we start at − 1? Could we make the numbers go down [decrease]?"

With the software that built lattices to the students' specifications, the children invented lattices as quickly as they could generate the ideas. They posed new problems:

- What is the difference between a lattice that starts at 3 and one that starts at 2?
- What happens if the lattice has only nine columns? Five columns? One column?
- What would the lattice look like that has ten columns and only even numbers?
- Could we create a ten-column lattice with the number 150 in the third row?
- Is it possible to make a lattice where halves are included?
- How many rows up would I need to go before reaching the number 542 in the ten-lattice? (This was easy to determine because the software permitted the children to move up many rows at a time.) How does this compare with the numbers of rows needed before reaching the number 542 in a twenty-column lattice?

Although the wording is ours, these were *student*-generated questions. Each new question inspired two or three new ones, and students were inventing and doing mathematics as they speculated on the answers. The freedom of the classroom led to a deeper understanding of the nature of number.

## WHY IS PROBLEM POSING GOOD FOR US?

An orientation toward *posing* new problems can be said to be the very heart of learning mathematics. Learning is a creative act: we learn not by absorbing but by *constructing* our knowledge. And we learn mathematics particularly well when we are actively engaged in creating not only the solution strategies but the problems that demand them.

What else is problem posing good for? It probably lessens mathematical anxiety. In certain classes where problem solving consisted of finding a solution to a teacher-posed problem, we noticed a great deal of anxiety among the students. There was a fear of being wrong or of thinking up foolish ideas. However, in a problem-posing environment, there is no one right answer. Students were willing to take risks, to pose what they considered to be interesting variations of the problem. As Brown and Walter (1983) point out mathematics became less "intimidating."

Problem posing also helps us to notice our misconceptions and preconceptions (see example in principle 3) and may, in that way, help us become better consumers in today's world. Presented thoughtfully, the problem-posing spirit can augment the rest of our teaching about the difference between the literal claims in advertising or political rhetoric ("it has three lines") and what we thought the product or promise really was (a triangle). By considering creative misinterpretations in the context of problem posing, we develop this new awareness.

Finally, problem posing helps to foster group learning rather than competition against the other members of the class.

In a world where the necessary mathematical skills are changing so rapidly, it will be the good problem poser who will be the best prepared.

## REFERENCES

Brown, S. I., & Walter, M. I. (1983). *The art of problem posing.* Hillsdale, NJ: Lawrence Erlbaum Associates.
Brown, S. I., & Walter, M. I. (1970). What if not? An elaboration and second illustration. *Mathematics Teaching, 51,* 9–17.

Goldenberg, E. P. (1970). Scrutinizing number charts. *Arithmetic Teacher, 17,* 645–53.
Greenes, C., Immerzeel, G., Okenga, E., Schulman, L., & Spungin, R. (1980). *Techniques of problem solving.* Palo Alto, CA: Dale Seymour Publications.
Lane County Mathematics Project (1981). *Problem solving in mathematics.* New Rochelle, NY: Cuisinaire Company of America. Author.
National Council of Teachers of Mathematics. (1989). *Curriculum and evaluation standards for school mathematics.* Reston VA: Author.
Ohio Department of Education (1982). *Problem solving: A basic mathematics goal.* Palo Alto, CA: Dale Seymour Publications.
Page, D. (1965). *Maneuvers on lattices.* Newton, MA: Education Development Center.

# FOOTNOTE

[1]Beyond merely citing the seminal writing of Stephen I. Brown and Marion I. Walter, the authors wish to acknowledge the personal influence these two teachers have had on our thinking and work.

# 20 Mathematical Investigation: Description, Rationale, and Example

Barry V. Kissane

A person given a fish is fed for a day.
A person taught to fish is fed for life.

Most areas of the curriculum experience some tension between content and process. History teachers are concerned with history as knowledge – what actually happened and why it happened – and history as activity – how do we know what happened and why? Our colleagues in science are concerned that children learn scientific information as well as the means by which it is acquired, commonly called the *scientific method*. Students of English learn about what has been written by others as well as how to write for themselves. So too does mathematics have both content and process dimensions, although it seems to me that we have long neglected the one in favor of the other. As John Dewey (1944) put it, "the record of knowledge, independent of its place as an outcome of inquiry and a resource in further inquiry, is taken to *be* knowledge."

*Mathematical investigation,* as conceived here, focuses on mathematics as something that people do rather than as something that people have done. I have in mind the metaphor of mathematics as a verb, as distinct from mathematics as a noun, recognizing that all metaphors have their dangers. I use the phrase to refer to the exploration of open situations in a relatively unstructured way. The intention of this article is to suggest what purposes might be served by such activity in the school mathematics curriculum and to offer an example of what is meant by it.

By its nature, of course, mathematical investigation is still concerned with mathematical content; but the focus is shifted toward the processes that are used to deal with the content. Some examples of processes relevant to mathematics and hence to mathematical investigation include posing problems, generating examples, specializing and generalizing, devising symbols and notation, recording observations, exploring a question systematically, identifying patterns, making and testing conjectures, justifying generalizations, and communicating with an audience. Some of these processes are components of what has lately been described as mathematical thinking; most are relevant to what many people seem to mean by problem solving, whereas several are related to what some think of as laboratory approaches to learning mathematics.

Although all these processes might be encountered within standard mathematics curricula, they are more likely to receive adequate treatment if explicit attention is given to them. Analogous to kissing, although it is conceivable that students will develop expertise by listening to how someone else does it, or even by watching others engaged in it, I doubt that an effective substitute can be found for personal experience.

A critical, defining feature of mathematical investigation is that the student is responsible for devising, refining, and pursuing the questions. It is not the teacher's or someone else's task. Why should mathematics consist entirely of answering questions that only someone else has asked? Is this approach not an essential distortion of the discipline?

## A RATIONALE

At least five reasons can be cited for reserving space in the curriculum for mathematical investigation:

1. It deals with the essential nature of mathematical activity—problem posing, dealing with situations that are not known in advance to have a single solution (or even any solution), and making and testing conjectures. The slick image of mathematics as a finished product should be exposed as the endpoint of much thinking rather than as the actual way in which mathematics is created.

2. It emphasizes those aspects of the discipline least susceptible to replacement by technology. We live in an age in which hand-held calculators to manipulate symbolic expressions, solve equations, analyze data, and

graph functions are already in production. Learning how to think mathematically will become more important in such an environment.

3. Persistence can be fostered by engaging students in a sustained way in a single task or set of tasks. Too much mathematics is presented to students in nibble-sized chunks, so that it is hardly surprising that they turn so readily to the teacher or the text for help.

4. Students will learn best about the nature of mathematics by actually doing it. Erroneous stereotypes, such as that mathematics is intrinsically difficult or that only brilliant people can do mathematics, can be addressed.

5. Perhaps motivation is the strongest rationale. Mathematical investigation furnishes a context in which students may actually care about what they are doing for its own sake. For too many students, most, if not all, motivation for learning mathematics seems at present to be external in origin. This fact is a sad indictment of our present curriculum.

Space is not adequate here to develop these points fully. Rather, I now offer a small example of a mathematical investigation in the hope that it illustrates the potential of this sort of activity.

## AN EXAMPLE

We begin with an open situation:
Investigate numbers adding to thirteen.

No explicit problem is formally stated here—no demand is made that students go in one direction rather than another. But that is not to say that mathematically rich problems are not lurking just under the surface. What might we do with this starting point? As a beginning, we might engage in some preliminary skirmishing, akin to mathematical doodling, to get a sense of what kinds of things seem interesting about numbers adding to thirteen. Some partitions of thirteen include the following:

$$1 + 12 \qquad 3 + 10 \qquad 6 + 7 \qquad 3 + 3 + 7$$
$$3 + 7 + 3 \qquad 2 + 2 + 2 + 2 + 5$$

A first question might occur to us:

*Question 1.*   How many partitions of thirteen are possible?
Systematically addressing this question will uncover a good deal of mathematics. How do we systematically address such a question? How do we know when we have all the partitions? Should we count $3 + 7 + 3$ and $3 + 3 + 7$ as one or as two partitions? Can we tell in advance how many partitions will occur? If we know the answer for thirteen, does that help us deal with fourteen? At this stage we have more questions than answers. But

what is so special about thirteen? A second question occurs to us, a "what-if-not . . . ?" question of the kind that Brown and Walter (1983) have so convincingly argued for because it is so often fruitful:

*Question 2.*   What if not thirteen? What if, say, fifteen?
Important components of mathematical thinking are specialization and generalization. To generalize, more than one case is needed. So we examine another case. What is the same? What is different? What is interesting? We see that we can also find many ways of writing fifteen as a sum:

$$4 + 4 + 7 \qquad 13 + 2 \qquad 13 + 1 + 1$$
$$7 + 8 \qquad 6 + 4 + 5 \qquad 2 + 3 + 4 + 6$$

Maybe knowing about partitions of thirteen would help us to find out about partitions of fifteen? We observe that both thirteen and fifteen can be written as sums of consecutive integers: $6 + 7$ and $7 + 8$, respectively. Another question occurs to us:

*Question 3.*   Which numbers can be expressed as sums of consecutive integers?
It looks rather straightforward at first: the odd numbers can be expressed as such a sum and the even numbers can't. How could we be sure about this statement? We examine some more cases and notice that fifteen can be written as the sum both of two and of three consecutive integers: $15 = 7 + 8$ and $15 = 4 + 5 + 6$. We can't do the same with thirteen. What is the reason for this difference? We then notice that some even integers can be written as sums, for example, $10 = 1 + 2 + 3 + 4$, but not as sums of two consecutive integers. Sums of three consecutive integers seem to be divisible by three, but sums of four consecutive integers are not divisible by four, or at least so it seems after a little more doodling. Why? Another question occurs to us:

*Question 4.*   What if not two consecutive integers? What if three? Or four?
Again, the "what-if-not" question is a fruitful one. Which numbers can be written as the sum both of two and of three consecutive integers? What do they have in common? Can we find numbers that can be written as the sum both of two and of five consecutive integers? Why? Why not? What about numbers that can be written as the sum of two, three, or four consecutive integers? A bigger question demands an answer here:

*Question 5.*   In how many different ways can a number be expressed as a sum of consecutive positive integers?

Can we predict the result for a particular number without writing out all the partitions? How do we know that we are correct? Conversely, what numbers can be written as sums of positive integers in exactly four different ways? What do they have in common? Do any at all exist? Do some numbers exist that cannot be written at all as the sum of consecutive positive integers? What else is special about them?

We have many questions here. Some are more interesting than others, some are bigger than others, and some are harder to answer than others. But all require mathematical thinking to come to terms with them, and all suggest themselves from an innocent-looking starting point. But we may yet go in other directions.

*Question 6.*  What-if-not addition? What if we multiply numbers together?

We again examine our initial partitions of thirteen from this fresh perspective:

$$1 \times 12 = 12 \qquad 3 \times 10 = 30 \qquad 5 \times 8 = 40$$
$$3 \times 3 \times 7 = 63 \qquad 2 \times 2 \times 2 \times 2 \times 5 = 80$$

Some products are bigger than other products. Why? Which of the products is the largest?

*Question 7.*  Which partitions give the greatest product?

To illustrate mathematical investigation more fully, we shall explore this question in greater detail than the previous ones. We begin by looking systematically at some products of the elements of partitions of thirteen:

$$3 \times 10 = 30 \qquad\qquad 4 \times 9 = 36$$
$$5 \times 8 = 40 \qquad\qquad 6 \times 7 = 42$$
$$3 \times 4 \times 6 \times = 72 \qquad 3 \times 5 \times 5 = 75$$
$$4 \times 4 \times 5 = 80$$
$$2 \times 4 \times 5 \times 2 = 80 \qquad 3 \times 3 \times 5 \times 2 = 90$$
$$3 \times 3 \times 3 \times 4 = 108$$

Being systematic bears some fruit. Some patterns begin to emerge. (Incidentally, why do we pay so much attention to helping students to identify patterns and so little attention to helping them to structure their work so that patterns will emerge for them to observe?) From some data of this kind, we can begin to make conjectures. The business of making conjectures, which then demands that we decide whether or not the conjectures are true, and if they are, why they are, is central to mathematical investigation. Two patterns are evident in the snapshot of data above. One conjecture, derived from reading the foregoing partitions from left to right, is the following:

*Conjecture 1.* Partitions for which numbers are closer together give larger products.

A second conjecture, derived from reading the foregoing partitions from top to bottom, is this:

*Conjecture 2.* Partitions with more elements give larger products than those with fewer elements.

To be systematic is to deal with these conjectures one at a time. It becomes useful at first and necessary later to state and deal with conjectures by using mathematical notation, here algebraic notation. Deciding on a means of symbolic representation is a crucial and oft-neglected part of elementary algebra. It is rare that students actually get to use algebra as a tool, so it is hardly surprising that they do not find it useful or interesting as often as we would like. We call a partition with $k$ elements a $k$-partition, and we represent it in square brackets. Thus a 2-partition of thirteen is [3, 10]. Dealing with conjecture 1 systematically is helped by first considering partitions into pairs of numbers. A little more data lead to a symbolic representation and formalization of our earlier informal observations:

*Conjecture 3(a).* The maximum product for 2-partitions of $2n + 1$ is $n(n + 1)$ from $[n, n + 1]$.    *(b).* The maximum product for 2-partitions of $2n$ is $n^2$ from $[n, n]$.

These conjectures can be tested against the available data, and more data can be generated to furnish a further test. A little algebra produces the mathematically convincing justifications, however. In conjecture 3*(b)*, if we partition $2n$ differently from $[n,n]$, say into $[n - k, n + k]$, the product is $n^2 - k^2$, which is clearly less than $n^2$. Such is the power of algebraic reasoning!

In further investigation of conjecture 1, we notice that we can pull ourselves up by our bootstraps—that the results for 2-partitions can be used to deal with larger partitions. For example, from the 3-partition [2, 5, 6], another 3-partition with a larger product can be derived by first thinking of the partition as [[2, 5], 6]. We can think of the [2, 5] as a partition of seven. But conjecture 3*(a)* suggests that [3, 4] yields a larger product than [2, 5], and indeed, it does. Thus we "improve" [2, 5, 6] to [3, 4, 6]. Continuing in this way, we improve the partition still further to [3, 5, 5] and then to [4, 4, 5], which seems to yield the maximum product (80) for a 3-partition of thirteen. This kind of thinking and some more examples lead us to make yet another conjecture:

*Conjecture 4.* The maximum product for a partition is obtained when no two numbers in the partition differ by more than one.

Again we can test our conjecture by generating some more data as well as by examining the data that already exist. Thus, we predict that for various-sized partitions of seventeen, the maximum products will come from [8, 9], [6, 6, 5], [4, 4, 4, 5], [3, 3, 3, 4, 4], [3, 3, 3, 3, 3, 2], and so on, and this prediction seems accurate; at least, we do not find counterexamples. These predictions and data also bear on conjecture 2, which we have so far neglected. The maximum products seem to get larger as we increase the size of the partition. For the partitions of seventeen here, the products in succession are 72, 180, 320, 432, and 486.

But the trend breaks down at that point. The maximal 7-partition of seventeen seems to be [3, 3, 3, 2, 2, 2, 2], for which the product of the terms is only 432, whereas the maximal 8-partition seems to be [3, 2, 2, 2, 2, 2, 2, 2], for which the product is only 384. Thus conjecture 2 is false, although exploring it was not unproductive. As often happens, a wrong conjecture is useful to us. We make progress because we made a conjecture. A little reflection persuades us that conjecture 2 must be false because the extreme case of the partition into *n* elements, each of which is unity, yields a product of only one! But this fact was not obvious at first. More data are needed to make a reasonable conjecture about the largest partition of a number. After some little time, making use of our existing results from conjectures 3 and 4, we obtain results like those in table 20.1.

A strong pattern seems to emerge. Successive numbers seem to be grouped in clumps of threes, with a predictable arrangement of elements in the maximal partition. We are encouraged to make the following further conjecture:

TABLE 20.1
Maximal Partition of Numbers

| Number | Maximal Partition |
|--------|-------------------|
| 7 | 3, 2, 2 |
| 8 | 3, 3, 2 |
| 9 | 3, 3, 3 |
| 10 | 3, 3, 2, 2 |
| 11 | 3, 3, 3, 2 |
| 12 | 3, 3, 3, 3 |
| 13 | 3, 3, 3, 2, 2 |
| 14 | 3, 3, 3, 3, 2 |
| 15 | 3, 3, 3, 3, 3 |

*Conjecture 6.* The maximum product is obtained when numbers are partitioned into twos and threes only in the following way:

For $3k$,
$$[3, 3, 3, \ldots, 3]$$
$$(k \text{ terms})$$
for $3k + 1$,
$$[3, 3, 3, \ldots, 3, 2, 2]$$
$$(k + 1 \text{ terms})$$
for $3k + 2$,
$$[3, 3, 3, \ldots, 3, 2]$$
$$(k + 2 \text{ terms})$$

It is not difficult to verify conjecture 6 informally, making use of the result suggested by conjecture 3. For example, for a number divisible by three, $3k$, the maximal products with $k - 1$, $k$, and $k + 1$ terms, respectively, come from the partitions

$$[3, 3, 3, 3, \ldots, 3, 4, 4, 4],$$
$$[3, 3, 3, 3, \ldots, 3, 3, 3, 3, 3],$$

and

$$[3, 3, 3, 3, \ldots, 3, 3, 3, 2, 2, 2].$$

The second of these is larger than the first because

$$3 \times 3 \times 3 \times 3 > 4 \times 4 \times 4.$$

The second is also larger than the third because $3 \times 3 > 2 \times 2 \times 2$. So the second partition — that partitioning $3k$ into all threes — gives the largest product.

What a curious result is suggested by conjecture 6 although we seem to have something of an answer to our question 7, we have at the same time generated yet another question, as happens frequently in mathematical investigation. Why should it be true that twos and threes together yield the maximum partitions? Stated very informally, we have the following question:

*Question 8.*    What is so special about twos and threes?
It is not clear how to address this intriguing question just yet, so we store it in the back of our mind for the moment. We are reminded, however, that mere data are insufficient for mathematical beliefs — we have to have reasons for believing things. Continuing in a spirit of exploration, it occurs to us that we have artificially restricted our search for maximal partitions to the partitioning of integers into integers. Although this approach may have

helped for a while, it may ultimately be unwise. What if not integers? Now we have a further question:

*Question 9.*    What happens with nonintegral partitions?

At first, the task of finding a maximal partition when the number of possibilities is infinite seems more than a little daunting, until we realize that we can use a continuous analog of conjecture 3 to get successive improvements to a partition:

$$[3, 3, 3, 4] \rightarrow [3, 3, 3.5, 3.5]$$
$$[3, 3, 3.5, 3.5] \rightarrow [3, 3.25, 3.25, 3.5]$$
$$[3, 3.25, 3.25, 3.5] \rightarrow [3.25, 3.25, 3.25, 3.25]$$

From such examples and such thinking, it is not difficult to arrive at the following conjecture:

*Conjecture 7.*    The maximum product occurs when all the elements of a partition are the same.

Although we do not yet have a formal proof of the conjecture, we are nonetheless convinced of its truth. We assume that we can construct a proof for the continuous case by finding that the maximum value of $(n - x)x$ occurs when $x = n/2$, but we don't stop to prove what we already believe. It is now a simple task on a scientific calculator to examine various partitions of thirteen to see which has the greatest product. See table 20.2.

We are convinced that the maximal partition of thirteen is that consisting of five equal terms, yielding a product of 118.81, to two decimal places. But how does this result help us more generally? Our question 9 has been refined to the following:

*Question 10.*    What size partition leads to the greatest product?

We examine more numbers using a calculator with the results shown in

TABLE 20.2
Products of Partitions of Thirteen

| Partition | Product (Rounded to Two Decimal Places) |
|---|---|
| [6.5, 6.5] | 42.25 |
| [13/3, 13/3, 13/3] | 81.37 |
| [3.25, 3.25, 3.25, 3.25] | 111.57 |
| **[2.6, 2.6, 2.6, 2.6, 2.6]** | **118.81** |
| [13/6, 13/6, 13/6, 13/6, 13/6, 13/6] | 103.46 |
| [13/7, 13/7, 13/7, 13/7, 13/7, 13/7, 13/7] | 76.19 |

table 20.3, in which $N$ is the number partitioned and $K$ is the number of equal elements in the partition of maximum product. The pattern is very clear and has some aesthetic appeal. We have no trouble making a reasonable conjecture about the general relationship:

*Conjecture 8.*    For integers between $3k + 1$ and $3k + 3$ inclusive, the maximum product occurs when the partition has $k + 1$ elements.

Our conjecture stands up against the data so far, but will it do so for larger numbers? We try some: the maximum for seventeen occurs with six elements, and that for twenty-five has nine. So far so good. But an empirical reason is not sufficient in mathematics. We need at least to test some more cases. At this point, the calculator becomes a little tedious to use, and we decide to write a short computer program to extend table 3 systematically. (How rarely do students write a computer program in mathematics to answer a question of real interest to them, for which they do not yet have the answer. Too often, programs are written to develop expertise at writing programs rather than with a real sense of purpose.) Because program 1 is not for widespread use and we will not store it for our own later use, it is stripped of the usual niceties of programming style and user-friendly input and output statements. The results appear in table 20.4.

Some surprises appear in table 4, marked by asterisks. It seems hard to believe that such a lovely pattern does not hold, and at first we do not believe it. Perhaps we have a round-off error in the computer or a bug in the program? We check by hand the first breakdown of the pattern, that for $N = 15$. We expected the maximal partition to be [3, 3, 3, 3, 3], which has a

TABLE 20.3
Size ($K$) of Maximal Partitions of Numbers ($N$)

| $N$ | $K$ |
|:---:|:---:|
| 1 | 1 |
| 2 | 1 |
| 3 | 1 |
| 4 | 2 |
| 5 | 2 |
| 6 | 2 |
| 7 | 3 |
| 8 | 3 |
| 9 | 3 |
| 10 | 4 |
| 11 | 4 |
| 12 | 4 |
| 13 | 5 |
| 14 | 5 |

| PROGRAM 1 |
|---|

```
10  FOR N = 1 TO 60
20      PROD = 0
30      FOR K = 1 TO N
40          IF (N/K)^K < PROD THEN GOTO 70
50          PROD = (N/K)^K
60      NEXT K
70      PRINT N, K - 1
80  NEXT N
```

TABLE 20.4
The Size of Maximal Partitions

| N | K | N | K | N | K | N | K |
|---|---|---|---|---|---|---|---|
| 1 | 1 | 16 | 6 | 31 | 11 | 46 | 17 |
| 2 | 1 | 17 | 6 | 32 | 12 | 47 | 17 |
| 3 | 1 | 18 | 7 | 33 | 12 | 48 | 18 |
| 4 | 2 | 19 | 7 | 34 | *13 | 49 | 18 |
| 5 | 2 | 20 | 7 | 35 | 13 | 50 | 18 |
| 6 | 2 | 21 | 8 | 36 | 13 | 51 | 19 |
| 7 | 3 | 22 | 8 | 37 | 14 | 52 | 19 |
| 8 | 3 | 23 | 8 | 38 | 14 | 53 | 19 |
| 9 | 3 | 24 | 9 | 39 | 14 | 54 | 20 |
| 10 | 4 | 25 | 9 | 40 | 15 | 55 | 20 |
| 11 | 4 | 26 | *10 | 41 | 15 | 56 | *21 |
| 12 | 4 | 27 | 10 | 42 | 15 | 57 | 21 |
| 13 | 5 | 28 | 10 | 43 | 16 | 58 | 21 |
| 14 | 5 | 29 | 11 | 44 | 16 | 59 | 22 |
| 15 | *6 | 30 | 11 | 45 | *17 | 60 | 22 |

product of $3^5 = 243$. Unfortunately, the computer is correct. The partition [2.5, 2.5, 2.5, 2.5, 2.5, 2.5] has a product of $(2.5)^6 \approx 244.1$; not much larger, but larger nonetheless. Piet Hein put it best: "Problems worthy of attack prove their worth by hitting back!"

Although conjecture 8 looked perfectly reasonable for the small numbers tested by hand, it breaks down for larger values. It seems to be tantalizingly consistent with much of the data, but of course the fact that it works most of the time is not sufficient for mathematics. This is the point of demanding reasons as well as data in mathematics. We should have an argument to support our beliefs, as mere induction from a pattern is a dangerous process. To make further progress with the problem, it may help to look at it from a slightly different perspective.

Rather than focusing on the size of the partitions, maybe we should look at the magnitude of the elements of the maximal partitions? Useful as it was, and frustratingly close as it was to solving the problem, perhaps question 10 was the wrong question to ask. Instead we consider the following:

*Question 11.*   What size elements lead to the maximum product?

Not much effort is required to change the computer program to print out the value of each element of the maximal partition. The easiest way is to change statement 70 to

70 PRINT N, K − 1, N/K − 1

and to run the program a second time. The results appear in table 20.5. Another kind of pattern emerges here—a strong sense of consistency, a sense of the convergence of results toward a constant value. It is not a difficult task to run the computer program for larger integers by changing statement 10 to test this conjecture informally with more data. Remarkably, we are encouraged to make the following conjecture:

*Conjecture 9.*   The maximum product occurs when each of the elements is as close as possible to *e*.

The ubiquitous $e = 2.718281828 \ldots$, the base of natural logarithms — that leprechaun of the number system named after the master Euler, has mysteriously appeared in yet another unexpected place! Conjecture 9 is consistent with the data in table 20.5. For example, if fifteen is partitioned into five, six, or seven equal elements, then each element is 3, 2.5, or 2.14, respectively. Of these, the partition into six elements, each 2.5, yields the greatest product. And, of the three numbers, 2.5 is the closest to *e*. We specialize again to check further. For $N = 56$, the partitions into twenty, twenty-one, and twenty-two equal elements yield elements of 2.80, 2.67, and 2.54, respectively. We find that the product is greatest with twenty-one elements and that 2.67 is the closest of the three numbers to *e*. The conjecture checks again. Conjecture 9 also sheds some light on question 8. Two and three are special because they are the integers closest to *e*. When partitions are composed of integers only, the average of the integers will be closest to *e* if all of them are either two or three! Mindful of the earlier difficulties with conjectures consistent with available data, we attempt this time to prove our result:

THEOREM. *The maximum value of $(x/n)^n$, for n a natural number, occurs when $(x/n)$ is as close as possible to e.*

A sketch of a proof might proceed as follows:
Let

$$P = \left(\frac{x}{n}\right)^n.$$

## TABLE 20.5
### The Elements of Maximal Partitions

| N | N/K | N | N/K | N | N/K | N | N/K |
|---|---|---|---|---|---|---|---|
| 1 | 1.00 | 16 | 2.67 | 31 | 2.82 | 46 | 2.71 |
| 2 | 2.00 | 17 | 2.83 | 32 | 2.67 | 47 | 2.77 |
| 3 | 3.00 | 18 | 2.57 | 33 | 2.75 | 48 | 2.67 |
| 4 | 2.00 | 19 | 2.71 | 34 | *2.62 | 49 | 2.72 |
| 5 | 2.50 | 20 | 2.85 | 35 | 2.70 | 50 | 2.78 |
| 6 | 3.00 | 21 | 2.63 | 36 | 2.77 | 51 | 2.68 |
| 7 | 2.33 | 22 | 2.75 | 37 | 2.64 | 52 | 2.74 |
| 8 | 2.67 | 23 | 2.87 | 38 | 2.71 | 53 | 2.79 |
| 9 | 3.00 | 24 | 2.66 | 39 | 2.79 | 54 | 2.70 |
| 10 | 2.50 | 25 | 2.77 | 40 | 2.67 | 55 | 2.75 |
| 11 | 2.75 | 26 | *2.60 | 41 | 2.73 | 56 | *2.67 |
| 12 | 3.00 | 27 | 2.70 | 42 | 2.80 | 57 | 2.71 |
| 13 | 2.60 | 28 | 2.80 | 43 | 2.69 | 58 | 2.76 |
| 14 | 2.80 | 29 | 2.63 | 44 | 2.75 | 59 | 2.68 |
| 15 | *2.50 | 30 | 2.73 | 45 | *2.65 | 60 | 2.73 |

Then

$$\ln P = n \ln \left(\frac{x}{n}\right).$$

(It is easier to find a maximum for ln $P$ than for $P$.) Take derivatives with respect to $n$:

$$(\ln P)' = 1 \cdot \ln \left(\frac{x}{n}\right) + (n)\left(-\frac{x}{n^2}\right)\left(\frac{n}{x}\right)$$

So

$$(\ln P)' = \ln \left(\frac{x}{n}\right) - 1.$$

At the maximum, $(\ln P)'$ is zero, so

$$\ln \left(\frac{x}{n}\right) = 1;$$

and so

$$\frac{x}{n} = e.$$

A more careful proof is needed to remove some of the loopholes, but we are persuaded that such a proof is possible and that conjecture 9 is true. We note too that the entire investigation might have been stated in the form of the foregoing theorem: "Prove that. . . ." I leave it to the reader to assess the costs and benefits of doing so. For my own part, I am more conscious of the costs.

## CONCLUSION

The example shows how mathematical investigation exposes and emphasizes mathematical thinking processes. Many situations can be used to engage students in this way—more than are commonly recognized. Bastow et al. (1984), for instance, offer forty of them—enough to keep any class going for more than a year, assuming that students are allocated a reasonable period of time to become involved with any investigation and to pursue it an appropriate distance. Although this article has not addressed questions of pedagogy or scheduling, these are dealt with in Bastow et al. (1984).

Engagement is the key. It is not enough to present students with mathematics, however elegant or attractive. The rationale sketched herein will be realized only if the students themselves are active participants—mathematical investigators.

# REFERENCES

Bastow, B., Hughes, J., Kissane, B., & Mortlock, R. (1984). *40 mathematical investigations.* Perth, Western Australia: Mathematical Association of Western Australia.

Brown, S. I., & Walter, M. I. (1983). *The art of problem posing.* Philadelphia: The Franklin Institute Press.

Dewey, J. (1944). *Democracy and education.* NY: Macmillan.

# 21 Curriculum Topics Through Problem Posing

Marion I. Walter

Problem *solving* has received much attention in the past years and younger children are beginning to be seriously involved in it. Problem *posing* is beginning to receive attention and it, too, can begin in the earlier years. Furthermore, if we encourage students to engage in problem posing, we can involve them more deeply in the development of topics that we wish *to cover;* in fact, we can use problem posing to help students *uncover* mathematics.

## ADDITION PROBLEMS

Students are often presented with pages of addition problems. What message does this give to the students? Surely one message is that they are not capable of making up their own practice problems. (I realise of course the advantage of giving students all the same problems – it is easier to check the answers!) What might one do instead and why?

Suppose the students are just beginning to learn how to add three digit numbers and that they have just worked out:

```
   3  4  2
 + 5  3  4
 ─────────
```

where there is no need to regroup or 'carry'. If we ask them to make up more such exercises, they will soon be faced with the challenge of what numbers to choose so that the total in each column is less than 10 if they wish to avoid having to regroup. Surely this would be a learning situation.

Or, suppose they are given:

```
  3  4  2
+ 2  1  8
```

Might it not be worthwhile to challenge the students to make up some more problems where the units column adds up to ten?

Or, let's turn the task around. Given:

```
  3  4  2
+ 5  3  4
```

what problems can you or your students pose?

One technique of problem posing just asks you to look at the given, in this case an addition problem, and asks you or your students to try to think of other problems. Brown and Walter (1983) have called this *accepting the given* and we sometimes call it *brute force* problem posing. Among the suggestions from participants at Lancaster, using only this technique of problem posing, were:

- Here the answer is 876. Make up other addition problems of two 3-digit numbers whose sum is 876. What do you think students will be discovering or learning if they do this?
- Make up other 3-digit addition problems whose answers consist of three consecutive digits.
- Rearrange the digits of each 3-digit number in 342 + 534 to get the largest possible total.
- Find all the different totals you can get by rearranging the digits.
- Make up a story that goes with 342 + 534.
- Make up other 3-digit problems for which the answer is such that the digit in the tens place is one greater than the digit in the units place.

Notice that each of these problems can raise or lead to other problems. For example, the second raises the additional problem: *what are all possible 3-digit numbers that consist of consecutive integers*? And what about the totals of 876 and 678? Can one always or ever reverse the digits of the addends to get the reversed total?

```
  3  1  2            2  1  3
+ 5  6  4   and    + 4  6  5
---------          ---------
  8  7  6            6  7  8
```

```
but   3  5  9        9  5  3      1  3  7      7  3  1
    + 5  1  7      + 7  1  5    + 7  3  9    + 9  3  7
    ---------      ---------    ---------    ---------
      8  7  6      1  6  6  8     8  7  6    1  6  6  8
```

If it is not possible, will one always get 1668 instead of 678? Explore.

The fourth problem might give rise to the question: *how many different totals are there*?

You will think of many other problems even without using any other techniques of problem posing. In this way students can be engaged in problem posing and problem solving while also getting practice in addition. Students will be thinking and will be involved in creating their own problems. Thus at an early age they can experience some mathematics in the making. They can learn from experience that mathematics is *not* a subject in which you have to be told everything and memorise a lot.

## Fractions

Next, I suggested a fraction excercise as a starting point: *what problems can you think of when faced with* $\frac{2}{3} + \frac{1}{5}$?

Usually students are given such problems, and they either have learned the algorithm for finding the answer or they make mistakes. They are usually not asked to think.

Among some interesting problems that were posed at Lancaster were these two:

- When in real life would you ever have to add these two fractions?
- How many different ways can you add these?

Some questions one might pose and which I imposed on the group are:

- Which is bigger, $\frac{2}{3}$ or $\frac{1}{5}$?
- Is the answer less or more than 1?
- By how much does $\frac{2}{3}$ differ from 1?
- By how much does $\frac{1}{5}$ differ from 1?
- What must one add to $\frac{2}{3} + \frac{1}{5}$ to obtain a total of 1?

I had worked this last question out and found:

$$\frac{2}{3} + \frac{1}{5} + \frac{2}{15} = 1.$$

Using Polya's admonition, which he so often said in his class: *Look at the problem,* I noticed that one could get the answer $\frac{2}{15}$ by multiplying the numerators and denominators of $\frac{2}{3}$ and $\frac{1}{5}$. I had chosen $\frac{2}{3} + \frac{1}{5}$ at random when I wrote it down on an overhead some time before Lancaster and had calculated the $\frac{2}{15}$. I was lucky! This immediately raised the question: *What* other fractions could one start with so that one could find the right answer by this 'wrong' method? This is such a rich problem that I have since 'milked' it a great deal. See Borasi (1986)—essay number 16 in this collection—for analysis of a similar problem in which wrong methods give right answers.


## Geometric Figure


Next I took a geometric starting point: a regular hexagon.

Instead of beginning with what one wants 'to teach' about a regular hexagon, (what does one want students to know about a regular hexagon anyway?), let us brute force problem pose. Here are some of the problems and questions that were suggested by the group (See figure 21.1):

• How many diagonals does it have?
• How many triangles can you make?


Note that this is a good example of a question that would need to be clarified by students. This is a valuable activity because it is misleading to always give clearly stated and well defined problems and learning to clarify problems is in itself a worthwhile activity.

• When all the diagonals are drawn in how many regions are formed?
• What is the length of each diagonal? Or one might ask: How many different lengths are there?
• What is the area of the hexagon formed by joining the alternate vertices?
• What is the largest circle you can draw in it?
• If it is not rigid, what shapes can you deform it into?


In Walter (1981) I suggested one of my favourite questions: *can you construct a regular hexagon, not only by using straightedge and compass, but from a paper circle, or a rectangle, or a scrap of paper or an equilateral triangle?*

If we draw diagrams suggested by some of the questions above, or other simple diagrams using a regular hexagon as a starting point, many more questions are suggested. Perhaps you want to pose a few now:

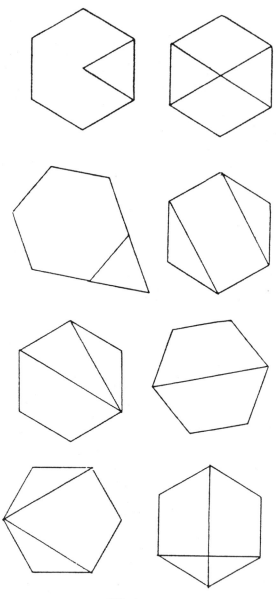

FIG. 21.1.

## Other Starting Points

We briefly examined $3x = 12$ as a starting point (Waler, 1988), as well as the problem of how many different triangles one could construct out of a stick of length 10 cm if each side has to be a whole number of centimeters and the stick gets used up each time one triangle is made. This is a nice problem that really brings home the triangle inequality — the sum of two sides of a triangle is greater than the third side. Warning: No easy formula exists that tells you how many triangles are possible for sticks of length n cm See Jordan (1979).

## Group Work

Part of the time was spent by participants in small groups choosing their own starting points — ones useful for the curriculum items they have 'to cover'. Since there were teachers of all levels from infant to upper secondary, we got a variety of topics, ranging from a delightfully rich one that started with the dustbin (figure 21.2) to a trigonometry one: $\sin^2 x + \cos^2 x = 1$.

FIG. 21.2.

The group discussing the dustbin created many wonderful questions including:

- How many dustbins are needed to collect the week's school litter?
- How high is it?
- What else can you measure about this dustbin?
- How many ways can you fit the lid on the bin?
- Is it better to have two little bins or one big one?
- What do you think would fit in this bin — an elephant? a bird . . .?
- Find the volume of the dustbin.
- If the bin is half full of water how high is the water level?
- How much bigger is the top rim than the bottom one?
- What else can you use if for? Would it be useful for storing toys?
- What if you could not reach into the bottom?
- What else can you put the rubbish in — what are some attributes of a good rubbish holder?

I believe the teacher who suggested this starting point had done a whole chunk of curriculum around the rubbish bin! After listing several questions suggested by $\sin^2 x + \cos^2 x = 1$, I pondered on how often we "teach" this without expressing the wonder about how special it is. Can we say anything about $\sin(x) + \cos(x)$, for example, or $\sin^2 x - \cos^2 x$ or $\tan(x) + \sin(x)$? I do not have a record of the work of the other groups at Lancaster, but I hope that even this small report will encourage others to engage in problem posing while trying to 'cover' the curriculum.

## REFERENCES

Borasi, R. (1986). Algebraic expressions of the error $16/64 = 1/4$. *Mathematics Teacher, 79,* 246-8.

Brown, S. I., & Walter, M. I. (1983). *The art of problem posing.* Hillsdale, NJ: Lawrence Erlbaum Associates.

Jordan, J. H. (1979). Triangles with integer sides. *American Mathematical Monthly, 86,* 686-689.

Walter, M. (1981). Do we rob students of a chance to learn? *For the Learning of Mathematics, 1*(3), 16-18.

Walter, M. (1988). Some roles of problem posing in the learning of mathematics. In D. Pimm (Ed.). *Mathematics, teachers and children* (pp. 190-200). London: Hodder and Stoughton.

# 22 Is the Graph of $y = kx$ Straight?

Alex Friedlander
Tommy Dreyfus

So as to create a little suspense, we declare in advance that the answer to the innocent question posed in the title depends on what is meant by $x$ and $y$. In this article, we shall show how the well-known fact that the graph of $y = kx$ in a Cartesian coordinate system *is* straight can be used as a point of departure for investigations into loci in non-Cartesian coordinate systems.

Geometric loci play a minor role in most school curricula, if only because they are not one of the items that are supposed to be in the bag of techniques known to the average high school graduate. They do, however, form an interesting and little-recognized way of looking at diverse concepts, which a student should learn. Many of the objects of Euclidean geometry can be viewed as loci, for example, the perpendicular bisector, the angle bisector, and the circle. Graphs of functions can also be viewed as loci. As a result, students have an opportunity to perceive them holistically as entities rather than as a collection of discrete preimage-image pairs of numbers. In this vein, we shall show that fairly complicated curves can be presented as loci at a quite elementary level.

The locus concept can be usefully presented throughout a student's mathematical development in a spiral sequence of topics at a gradually increasing level of formality and abstractness. Traditionally, loci have been introduced in the following sequence:

1. In the lower grades, pupils study geometrical objects and their properties. Circles, angle bisectors, and perpendicular bisectors, together with their intuitive locus properties, are presented at this stage.

2. The formal locus concept appears toward the end of a Euclidean geometry course. At this point, the previously encountered objects are defined as geometric loci.
3. The conic sections and other loci are presented in algebra and analytic geometry courses in a formal and mainly algebraic manner.

This traditional approach presents two difficulties. As a result of the general neglect of geometry at the elementary school level, and as a result of the fact that loci appear late – if at all – in Euclidean geometry courses, the geometric stages 1 and 2 are frequently curtailed or even omitted. Moreover, even if stages 1 and 2 are completed, a considerable gap remains between stage 2, when only the most elementary loci are presented, and stage 3, at which the topic receives a rather abstract, predominantly algebraic treatment. In other words, most students reach stage 3 without the appropriate preparation, and the notion of geometric locus becomes another topic in which the main activity is formal manipulation. At this stage, students have lost the opportunity to grasp loci intuitively as global entities that would be part of an integrated view of mathematics.

## THE IDEA

The omission of an intuitive, visual, geometrical stage in the introduction of more complex loci is not the result of oversight. With few exceptions, the task of locating a sufficient number of discrete points belonging to a certain locus to enable the student to sketch the corresponding graph in a Cartesian coordinate system is a technically difficult and aesthetically unrewarding task. Take, for example, the relatively simple locus of all points that are equidistant from a given line and from a given point (i.e., a parabola). Constructing the points is a lengthy and messy process of drawing parallels to the given line and circles around the given point and marking corresponding intersection points. One way out is to exploit computer graphics. Another way is the use of an appropriate coordinate system, for example, the focus-directrix system (Rose 1974). In this unusual coordinate system (fig. 22.1B), the parabola mentioned previously is represented by the surprisingly simple relationship $y = x$, where the $x$- and $y$-coordinates represent the distances from the directrix (the given line) and the focus (the given point), respectively. Similarly, ellipses can be represented in the focus-directrix system by a linear relationship $y = kx$, $0 < k < 1$ (fig. 22.2B).

In the following sequence of activities we shall construct and investigate a variety of loci that are all based on a ratio relationship $y = kx$. For every problem, we make a convenient, natural choice of coordinates and supply

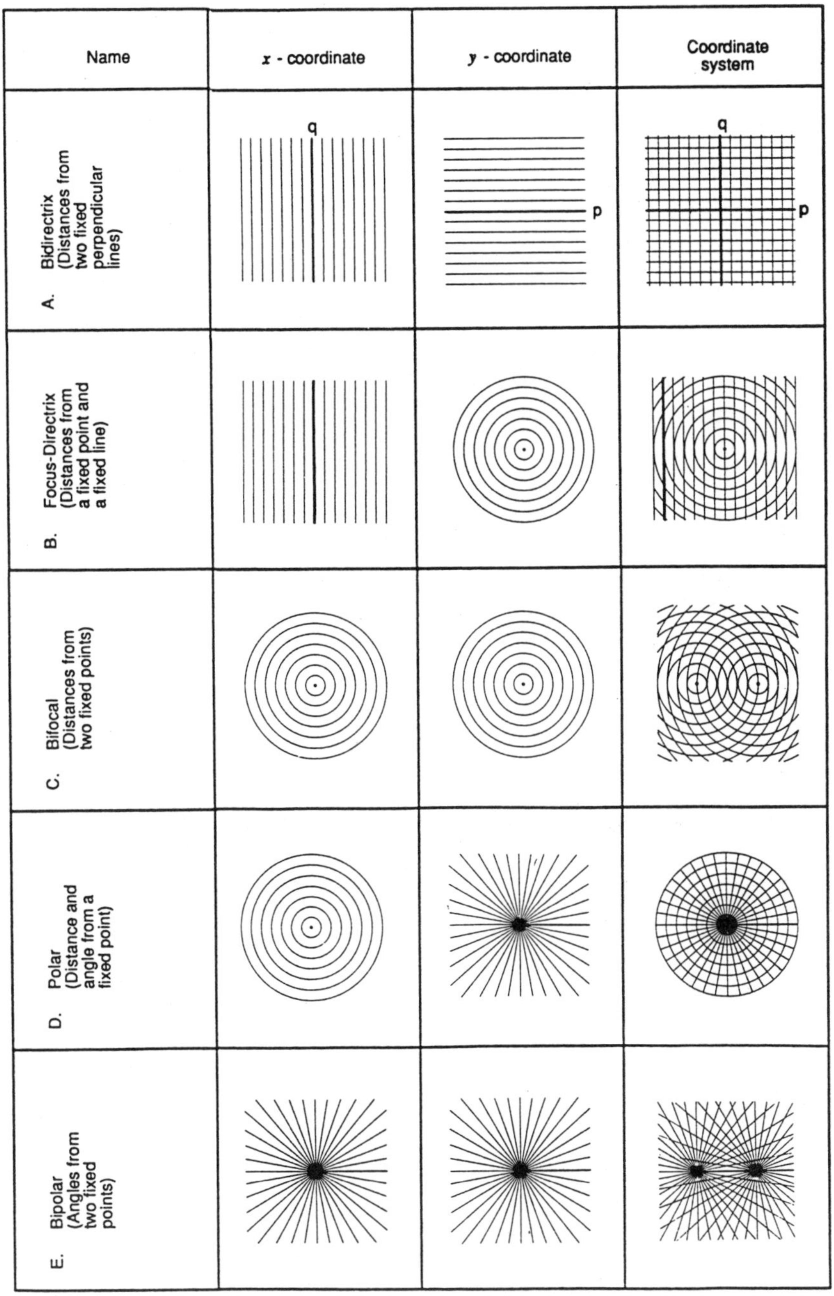

| Name | $x$ - coordinate | $y$ - coordinate | Coordinate system |
|---|---|---|---|
| A. Bidirectrix (Distances from two fixed perpendicular lines) | | | |
| B. Focus-Directrix (Distances from a fixed point and a fixed line) | | | |
| C. Bifocal (Distances from two fixed points) | | | |
| D. Polar (Distance and angle from a fixed point) | | | |
| E. Bipolar (Angles from two fixed points) | | | |

FIG. 22.1.   Coordinate systems

213

corresponding ready-made coordinate sheets that teachers can duplicate for students. The activities described here have been used with students, preservice teachers, and in-service teachers.

## GETTING STARTED: THE BIDIRECTRIX SYSTEM

*Given two perpendicular lines p and q, find the locus of all points whose distance from p is twice their distance from q.*

The natural way to find this locus is to build a coordinate system based on the two given lines; accordingly, we shall call the lines *axes* or *directrices,* and the system, the *bidirectrix system.* With an eye to the following activities, we are very careful and explicit with respect to the construction of the coordinate system and the meaning of the x- and y-coordinates. That is, we stress explicitly that in this case, the x-coordinate of a point is its distance from the line q, and the y-coordinate of a point is its distance from the line p. To measure these distances it is convenient to introduce parallels to the axes, that is, to work with a square grid (fig. 22.1A). Such a grid enables us to locate easily several points that satisfy the required relationship $y = 2x$, in which the ratio $y/x$ between the distances equals 2. A consideration of one or two points with nonintegral coordinates gives the required interpolation between the grid points. Thus, we observe that the locus is composed of a pair of straight lines (fig. 22.2A). We further graph some more loci with different distance ratios, that is, different positive values of $k$, including $k = 1$. We also discuss the relationship between the value of $k$ and the slopes of the lines. In class, this discussion can be enhanced by using an overhead projector or a computer screen, on which two lines move to various positions on a prepared bidirectrix system.

Clearly, the bidirectrix system is closely related to the Cartesian coordinate system, and the two can be compared. We stress the fact that the two systems are not identical: the distances from the directrices $p$ and $q$ are always positive and correspond to the absolute value of the Cartesian x-y coordinates. Thus any given pair of positive numbers will be represented in the bidirectrix system by four symmetric points, whereas it is represented by a single point in the Cartesian system. The relationship $y = kx$ $(k > 0)$ is represented by a pair of lines in the bidirectrix system, whereas in the Cartesian system it is represented by a single line.

## SOME OTHER COORDINATE SYSTEMS

*What changes do we need to introduce if instead of two perpendicular lines we are given a point and a line and are looking for the locus of all points whose distance from the point is twice their distance from the line?*

The natural way to find the locus is to build a coordinate system based on the two given elements—the point, which will be called the *focus,* and the line, the *directrix;* accordingly, one speaks of the *focus-directrix system.* Similarly to the previous case, we measure distances from the line by using a system of parallel lines. By contrast, distances from the focus are conveniently measured by using a system of concentric circles around the given point (fig. 22.1B). Before graphing the required locus, students may need some practice to get acquainted with the correspondence between pairs of positive numbers and points on the focus-directrix grid, that is, to see this new grid as a coordinate system. It should be noted explicitly that the required locus, a hyperbola, is described by the equation $y = 2x$, where $x$ denotes the distance from the directrix and $y$, the distance from the focus.

Students should now be asked to investigate the qualitative changes the locus undergoes when the ratio between the distances is changed, that is, to consider the graphical representation of $y = kx$ for other values of $k$. Thus they learn that for ratios $k < 1$, the locus is an ellipse; for $k = 1$, a parabola; and for $k > 1$, a hyperbola. This intuitive introduction presents a unified description of all three conic sections and shows the role of the eccentricity coefficient $k$ in determining the graph (fig. 22.2B).

A similar approach can be taken for the *bifocal system,* where the given elements are two points and the task is to find the locus of all points for which the distance from one of the given points is a fixed multiple of the distance from the other. The grid is formed by two systems of concentric circles (fig. 22.1C). Again, a correspondence occurs between pairs of positive numbers and points in the grid, and by using this correspondence, the required loci can easily be graphed for various values of the ratio (fig. 22.2C). Again, the algebraic relationship is $y = kx$, where $x$ and $y$ are the distances from the two foci, respectively, and, except for the well-known case of the perpendicular bisector ($k = 1$), all the loci turn out to be circles. It is interesting to note that a symmetry between the loci for inverse values of $k$, such as 2 and 1/2, naturally occurs in the bidirectrix and in the bifocal system but not in the focus-directrix system because of the different nature of the given elements.

The well-known polar coordinate system suggests the use of coordinate systems in which at least one of the coordinates is an angle. The angle is measured from a given direction at a given point, the pole. The other coordinate can be a second angle or a distance. If the distance from the pole is chosen as second coordinate, the *polar coordinate system* itself results. The appropriate grid consists of a set of concentric circles around the pole and a set of rays emanating from the pole (fig. 22.1D). For convenience, we choose the unit angle to be 10 degrees and let the angle vary over all positive numbers, excluding zero to avoid ambiguities. As before, a correspondence exists between pairs of positive numbers and points in the grid, and by using this correspondence, the required loci can easily be graphed for various values of the ratio (fig. 22.2D). After the graphing activity, the algebraic relationship $y = kx$ can be reemphasized, wherein $x$ is the angle and $y$ is the distance from the pole. All loci in this case are Archimedean spirals. Different values of the ratio $k$ will change the "speed of rotation" but will not cause qualitative changes of the loci.

If both coordinates are chosen to be angles, two poles are needed. The resulting coordinate system will therefore be called the *bipolar system*. The appropriate grid consists of two sets of rays, one at each pole (fig. 22.1E). By using this grid, the loci of points such that one of the angles is a constant multiple of the other are just as easy to graph as the previous loci. Their shape does, however, vary widely, depending on the value of the ratio. The easiest cases are the following: a ratio of 1 yields an empty locus because equal angles have parallel sides, a ratio of 2 or 1/2 yields circles, and a ratio of 3 or 1/3 yields right strophoids (fig. 22.2E). Note that the two circles, as well as the two strophoids, are symmetric to each other with respect to the perpendicular bisector of the line joining the poles; as in the bidirectrix and in the bifocal system, the reason for this occurrence is the symmetry between the two given elements.

As is evident from figure 22.2, in the process of building loci, students meet some previously known lines and curves (e.g., the perpendicular bisector and circles in C) and also discover some new graphs (conic sections in B, Archimedean spirals in D, strophoids in E). The possibility of an empty locus arises for $y = x$ in the bipolar system. The encounter with already known loci presents an opportunity to discuss alternative definitions for the same curve. The circle, for example, can also be defined as the locus of all points whose distances from two fixed points have a constant ratio not equal to 1. In the process of new discoveries, the corresponding loci can be discussed and named. It would seem appropriate to the approach that these discussions should emphasize geometric properties rather than analytic aspects of the loci. Comparisons of the degree of "openness" of a parabola with that of a hyperbola or with that of self-intersecting curves demonstrate such properties.

## LOCI AND THE "WHAT-IF-NOT" STRATEGY

The activities suggested in this paper basically take the problem of graphing $y = kx$ in a Cartesian coordinate system and apply the "what-if-not" problem-posing strategy described by Brown and Walter (1983). As generally pointed out in their book, an almost endless list of problems can result

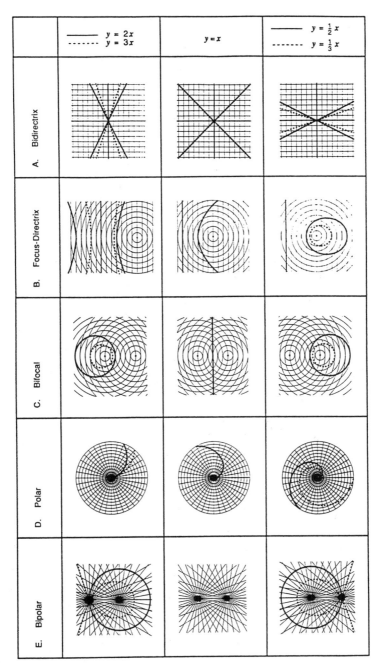

FIG. 22.2.   Loci in coordinate systems

by negating different attributes of a given problem. Most of the investigations proposed here are a result of negating the Cartesian attribute of the original problem.

We suggest next a few more paths of inquiry of the "what-if-not" type that can be used to enhance our original purpose of filling the gap between the elementary aspects and the algebraic treatment of geometric loci.

• *What if the distance between the foci or between focus and directrix is different or variable*? Teachers can use the fact that the distance between the fixed elements does not alter the type of locus to facilitate the sketching of certain graphs. In the focus-directrix system, a distance of 6 units allows whole-number coordinates for the vertices related to $y = 2x$ or $y = \frac{1}{2}x$, whereas a distance of 8 is more convenient for the parameters 3 and 1/3. Weaker students have been observed to miss the vertices altogether if their coordinates are not integral.

• *What if the pole is not the focus in the polar coordinate system*? That is, what is the locus of all points whose distance from one given point – the focus – is a multiple of the angle at another given point – the pole? The treatment of this case will be based on yet another coordinate system, the *focus-pole system*.

• *What if the angle between the axes in the bidirectrix system is changed so that the axes are oblique or even parallel*? The latter case is discussed in an article by Friedlander, Rosen, and Bruckheimer (1982).

• *In the bidirectrix system, what if one changes the Euclidean definition of distance,*

$$d = \sqrt{x^2 + y^2},$$

*to the "taxicab" distance $d = |x| + |y|$?* The graphs that result have been discussed by Krause (1973).

• *What if we choose the distance from a line and an angle at a point as coordinates*? This investigation leads to the only remaining coordinate system that can be built with distances from a point, distances from a line, and angles as coordinates; it is called the *pole-directrix system*. It turns out that the graph of $y = kx$ in this system is not a simple curve but consists of an infinity of branches, somewhat like the graph of $y = \tan x$ in a Cartesian coordinate system. In fact, these loci are related to the tangent function: the curve whose equation in the pole-directrix system is $y = kx$ is described in a Cartesian system by $y = x \tan (kx)$.

The aforementioned ideas show that the use of the "what-if-not" strategy leads to a great deal of flexibility and allows one to extend considerably the scope of the mathematics covered, from a simple proportionality relationship $y = kx$ all the way to conic sections and strophoids.

## REFERENCES

Brown, S. I., & Walter, M. I. (1983). *The art of problem posing.* Philadelphia: The Franklin Institute Press.

Friedlander, A., Rosen, G., & Bruckheimer, M. (1982). Parallel Coordinate Axes. *Mathematics Teaching, 99,* 44–48.

Krause, E. F. (1973). Taxicab Geometry. *Mathematics Teacher, 66,* 695–706.

Rose, K. (1974). New Conic Graph Paper. *Mathematics Teacher, 67,* 604–606.

# Your Turn
# Editors' Comments

Anything such as a situation, problem, theorem, or data, for example, can be enriched by posing new problems. You now might want to try your hand at problem posing using an arithmetic situation as a starting point. Consider the well known hotel problem:

> There are n hotel rooms, numbered from 1 to n, on a long corridor. Guest 1 opens all the doors. Guest 2 closes every second door starting at door 2. Guest 3 changed the position of every third door starting with door 3 and so on for n guests.

Which doors are left open at the end of this process is the usual question asked. What problems can you pose? After exploring the possibilities you may want to read **Because a Door Has to be Open or Closed** in which Cassidy and Hodgson discuss some of their interesting variations using the hotel problem as a starting point. As you compare your work with theirs, you will see that everyone has a chance to come up with different ideas. They rightly warn or remind us of the fact that the What-if-Not technique "may generate a new problem that is most difficult to solve." They stress that a new problem generated in a problem posing spirit may be "absurd, trivial or even 'impossible' to solve". We fully agree and feel that learning to sense that this might be the case, is often very difficult but is part of becoming a good problem poser and solver. Their article and the techniques they illustrate should serve as an example to encourage readers to problem pose on their own chosen situations and to invite their students to do the same.

# 23

# Because A Door Has To Be Open Or Closed: An Intriguing Problem Solved By Some Inductive Exploration[1]

Charles Cassidy
Bernard R. Hodgson

Much too often, students will have a mostly passive attitude when they confront mathematics in school: they wait for the mathematical situation to be created for them, they rely on the teacher to formulate the problems to be solved, they expect to be given indications about the right theorem to be used in the right place, and so on. This is not surprising, since mathematics is too often presented as a "polished" subject with almost no place for personal contributions.

One way for teachers to stimulate students' involvement is by encouraging them to generate problems of their own. Such a *problem-posing approach* has been studied extensively by Walter and Brown (1977), particularly in connection with the what-if-not? strategy. Roughly speaking, this technique may be described as choosing one attribute of a phenomenon, varying that attribute, posing a question about the varied attribute, and trying to solve the new problem thus created.

We shall now use the what-if-not? technique to transform a well-known problem into a new one. Our purpose is twofold: to illustrate this technique and to bring to the reader a problem that we found to be of considerable interest in its own right.

### THE "WHAT-IF-NOT?"

We start with the following "door problem," which seems to be a classical one in the teacher-training literature (see, for instance, the versions in Bell,

Fuson, and Lesh [1976, p. 620] or Leblanc, Kerr, and Thompson [1976, p. 79]).

## Problem 1

At Long Hotel in Long Beach, there are $n$ rooms all located along a very long corridor and numbered consecutively from 1 to $n$. One night after a party, $n$ people, who have been likewise numbered from 1 to $n$ (in order to hide their true identities), arrived at this hotel and proceeded as follows: Guest 1 opened all the doors. Then Guest 2 closed every second door beginning with door 2. Afterwards, Guest 3 *changed the position* of every third door starting with the third door (that is, the guest opened the doors that were closed and closed those that were open). In a similar way, Guest 4 changed the position of doors 4, 8, 12, . . . . This process continued until each person had walked the length of the corridor. Of course, the last person, Guest $n$, merely strolled to the end of the corridor, where the guest changed the position of door $n$. We pose the following question: *Which doors were left open and which ones were left closed at the end of the process?*

At first glance this problem seems quite complicated. Yet, just a little thought reveals the following underlying principle: At the end of the process, door $d$ will be open or closed depending only on the parity of the number of divisors of $d$. For example, assuming $n \geq 10$, we note that door 9 will be touched by Guests 1, 3, and 9, and hence will end up open. Door 10 will be touched by Guests 1, 2, 5, and 10, and hence will end up closed. Now, it is the case that the number of divisors of a given number $d$ will be odd if and only if $d$ is a perfect square. (This is not hard to prove; think about it.) *Thus, all doors will end up closed except those whose numbers are perfect squares.* In view of the apparent "complexity" of the problem, this answer is remarkably simple. Indeed, it is this contrast that gives real interest to the problem.

We now apply the what-if-not? technique to the preceding problem. Drawn by the fact that the rooms are along a straight corridor, we ask: "*What if* the corridor is *not* straight, but rather is circular? Well, it might be said, guests would possibly have the idea of going around more than once. But then when would they stop?" We are thus led to stipulate new rules in order to determine the exact behavior of the guests. So, we pose the following problem, which provides us with a discovery situation undoubtedly still easy enough to be worked by our students.

## Problem 2

The rooms of Circle Hotel, above the Polar Circle, were built around the perimeter of a circular courtyard and were numbered consecutively from 1 to $n$. Owing to bad weather conditions in California, the same $n$ persons encountered in problem 1 were spending their winter holidays at Circle Hotel. One night, again, after a party, they tried to repeat what they had done before at Long Hotel. However, something was different. Since the hotel had the form of a circle, they could have opened and shut the doors indefinitely. So, the manager asked a watchman to stop each one as soon as he or she had changed the position of door $n$. (Fortunately, this always happened even if for some it took a long time.)

What actually took place is the following: Guest 1 opened all the doors and was stopped as soon as he opened door $n$. Then Guest 2 closed every second door beginning with door 2; but since $n$ may be either odd or even, all we can say is that the guest was stopped after one revolution around the hotel if $n$ was even or at the end of two revolutions if $n$ was odd. Afterwards, Guest 3 changed the position of every third door starting with door 3 and was stopped on opening (or shutting) door $n$. This process continued until Guest $n$ had finished. We pose again the same question: *Which doors were left open and which ones were left closed at the end of the process?* The reader might now like to try solving problem 2 before continuing.

*Solving problem 2.* Our problem is of the sort that responds to an *inductive approach.* After experimenting with some concrete cases, we look for a pattern; that is, we "guess" what the general rule should be and then try to justify that conjecture.

The results obtained by considering seven different hotels with, respectively, two to eight rooms are given in table 23.1.

It appears that whatever the number of rooms in the hotel, *only one door is left open.* Moreover, we notice, at least for these small values of $n$, that *if $n$ is odd, then door $n$ is the one left open; and if $n$ is even, then door $n/2$ is the one left open.* It is quite natural to conjecture that this is the solution to problem 2.

Now consider the $(n - 1)$ doors that (apparently) end up closed. There

TABLE 23.1

| Total Number of Rooms in the Hotel | The Doors That Will Be Left Open | Those That Will Be Closed |
|---|---|---|
| 2 | 1 | 2 |
| 3 | 3 | 1, 2 |
| 4 | 2 | all others |
| 5 | 5 | all others |
| 6 | 3 | all others |
| 7 | 7 | all others |
| 8 | 4 | all others |

must be an *even* number of guests who change the position of each one of these doors. Taking the set of persons who do change the position of a given door, we are thus tempted to try to subdivide it into subsets, each containing two people, and to do this subdivision in a meaningful manner. This subproblem also can be approached inductively through experimentation with hotels having a small number of rooms. It turns out that the case $n = 12$ is particularly instructive. In table 23.2, we put an x at the intersection of line $i$ and column $j$ whenever Guest $i$ happens to change the position of door $j$.

A striking symmetry is seen in this table. If line 6 is thought as an "axis of reflection," the upper half of the table is reflected in the bottom half (except that line 12 has no symmetric counterpart). This means that Guest 1 and Guest 11 touch the same doors, as do Guests 2 and 10, 3 and 9, 4 and 8, and 5 and 7. We thus have a clue about the subdivision that we are looking for and conjecture at this point that *in a hotel with n rooms, if Guest k (k ≠ n) changes the position of a given door, then so does Guest (n − k)*.

It is not difficult to justify this conjecture. Guest $k$ is stopped as soon as he changes the position of door $n$; this means, in particular, that he has just previously touched door $(n - k)$ and, before that, door $(n - 2k)(\bmod n)$, door $(n - 3k)(\bmod n)$, and so on; whereas Guest $(n - k)$ touches consecutively door $(n - k)$, door $2(n - k)(\bmod n)$, door $3(n - k)(\bmod n)$, and so on. Hence the assertion is verified. For example, in the case of the hotel with 105 rooms, Guest 30 touches consecutively doors whose numbers are 30, 60, 90, 15, 45, 75, and 105, whereas Guest 75 touches consecutively doors 75, 45, 15, 90, 60, 30, and 105.

TABLE 23.2

| | | | | | Room | Number | | | | | | |
|---|---|---|---|---|---|---|---|---|---|---|---|---|
| | 1 | 2 | 3 | 4 | 5 | 6 | 7 | 8 | 9 | 10 | 11 | 12 |
| 1 | x | x | x | x | x | x | x | x | x | x | x | x |
| 2 | | x | | x | | x | | x | | x | | x |
| 3 | | | x | | | x | | | x | | | x |
| 4 | | | | x | | | | x | | | | x |
| 5 | x | x | x | x | x | x | x | x | x | x | x | x |
| 6 | | | | | | x | | | | | | x |
| 7 | x | x | x | x | x | x | x | x | x | x | x | x |
| 8 | | | | x | | | | x | | | | x |
| 9 | | | x | | | x | | | x | | | x |
| 10 | | x | | x | | x | | x | | x | | x |
| 11 | x | x | x | x | x | x | x | x | x | x | x | x |
| 12 | | | | | | | | | | | | x |

(Guest Number is labeled along the left side of the rows.)

Our problem is now almost solved. Indeed, we may readily conclude from the discussion above that *most doors* are touched an even number of times, and thus end up closed. What are the exceptions? To answer this question, we must take a closer look at our now proven conjecture to see how exceptional cases can arise.

First, it is clear that for any $n$, Guest $k$ has no "symmetric partner" when $k = n$: indeed, since there is no Guest 0, Guest $(n - k)$ simply does not exist in such a case. But this does not mean that door $n$ will always be left open. To see why this is so, let us introduce the second exceptional case: If $n$ is even and $k = n/2$, then Guest $k$ is the same person as Guest $(n - k)$.

The problem thus reduces to considering the action of Guest $n$ if $n$ is odd and to considering the action of both Guest $n/2$ and Guest $n$ in the case where $n$ is even. It is now clear that if $n$ is odd, door $n$ will be left open and the other ones will be left closed; whereas if $n$ is even, door $n/2$ will be left open and all the other ones will be left closed.

## RELATION TO THE SPIROGRAPH®

We would like to mention briefly that there are many combinatorial questions that arise naturally in connection with problem 2. For example, how many times does Guest $k$ go around the hotel before being stopped by the watchman? How many doors has the guest then opened or closed? Such questions also arise in studying the Spirograph®, and it is interesting to stress the analogy between both situations.

For instance, suppose that the circular hotel has 105 doors and consider the behavior of Guest 30, who touches consecutively the doors with numbers 30, 60, 90, 15, 45, 75, and 105 (note that in such a hotel, door 15 is 30 doors after door 90). Then consider the design produced with the Spirograph® by rolling the small wheel of 30 teeth inside the big wheel of 105 teeth; the seven-pointed star obtained is shown in figure 23.1. The seven points can thus be interpreted as falling, in order, opposite the teeth with numbers 30, 60, 90, 15, 45, 75, and 105 of the big wheel.

More generally the behavior of Guest $k$ in a hotel with $n$ doors is similar to that of a small wheel of $k$ teeth rolling inside a big wheel of $n$ teeth. (It is quite clear, however, that the *circular-hotel-Spirograph®* analogy is not perfect: the commercial Spirograph® provides wheels for only a few values of $k$ and $n$, and, furthermore, certain values of $k$ or $n$ are physically impossible.)

The principle underlying the Spirograph® has already been studied in Cavanaugh (1976), so we refer the reader to this paper for full details. Using

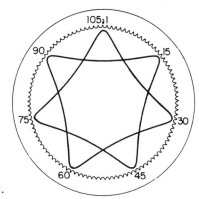

FIG. 23.1.

the analogy mentioned above, we find the following information about Guest $k$'s door-changing activity in a hotel with $n$ rooms. If $m$ and $g$ denote, respectively, the L.C.M. and the G.C.D. of $n$ and $k$, then the number $r$ of revolutions of Guest $k$ around the hotel is given by

$$r = \frac{m}{n} = \frac{k}{g};$$

and the number $s$ of doors whose position the guest has then changed (which also corresponds to the number of points of the design obtained by the Spirograph®) is given by

$$s = \frac{m}{k} = \frac{n}{g}.$$

## CONCLUSION

We hope that the preceding discussion about problem 2 serves as a good illustration both of the problem-posing attitude that the teacher should have and of the powerfulness of the inductive approach in analyzing a problem that looks complicated at first glance.

However, it should be mentioned that there is an inherent danger in the what-if-not? strategy, namely, that we may thus generate a new problem that is most difficult to solve. This aspect cannot be ignored, but in fact we never really know how difficult a problem is before we try it. Also, it might well be that the problem obtained is absurd, trivial, or even "impossible" to solve. It is hoped that experience will provide the teacher with a feeling for which what-if-not? questions will lead to "good" problems.

# REFERENCES

Bell, M. S., Fuson, K. C., & Lesh, R. A. (1976). *Algebraic and arithmetic structures.* NY: The Free Press.

Cavanaugh, W. E. (1975). The Spirograph and the greatest common factor. *Mathematics Teacher, 68,* 162-163.

Leblanc, J. F., Kerr, D. R. Jr., & Thompson, M. (1976). *Number Theory* (Mathematics Methods Program). Reading, MA: Addison-Wesley.

Walter, M. I., & Brown, S. I. (1977). Problem posing and problem solving: An illustration of their interdependence. *Mathematics Teacher, 70,* 4-13.

# NOTE

1. The authors express their gratitude to Graham Lord for his help in the preparation of the original manuscript and also the referees of *The Mathematics Teacher* whose most valuable remarks greatly improved the final presentation of the text.

# III Geometry
# Editors' Comment

As you probably noticed in reading the previous chapter, the demarcation between algebraic and geometric concepts is not so sharply divided. What appears to be an algebraic idea is often translatable in geometric terms and vice versa. As a matter of fact, one very helpful heuristic for generating problems is to ask how it is that what appears to be essentially algebraic (or geometric) might be viewed geometrically (algebraically). The idea is helpful not only for the purpose of generating new problems but of understanding ones that we may be engaged in solving as well. Among the questions we have listed in the section entitled, *A Handy List of Questions* in **The Art of Problem Posing** are:

How can you view it geometrically?
How can you view it algebraically? (p. 27, 28)

You may wish to read that section since it suggests a number of interesting problem posing strategies that do not require a What-if-Not point of view (not at least in a direct sense).

Though we found it helpful to partition chapters II and III according to their algebraic and geometric content, we hope you will keep the delicate reverberation between the two content areas in mind as you read this chapter, as you re-read the other chapters in the text and as you attempt to generate new problems on your own.

This chapter is organized into four clusters. The first (including articles by Sowder, Jones and Brown) explores problem generation from the perspective of **looking back** after one has solved a problem. The second (including articles by Chazan and Hoehn) generalizes the substance of the first cluster by exploring a number of ways in which **the intertwine of problem posing and problem solving** operate. The third section (including excerpts of two articles by Walter), **something comes from nothing** is a play on a popular song from the movie version of The Sound of Music. Walter's articles show how the *ordinary* world around us has within it the potential to encourage us to raise some fascinating questions. In the fourth section (which includes the article by Schmidt) entitled **your turn,** we encourage you to do what you did at the end of chapter II—to employ the strategies of problem posing on a new situation before actually reading the article.

# Looking Back
# Editors' Comments

A well known but little practiced problem solving heuristic that is included among those that Polya (1962, 1973) has spoken of so eloquently is that of *looking back* after one has solved a problem. It is by looking back that one may become explicitly aware of positive (and also negative) strategies that have been used but perhaps not incorporated into one's awareness. In the act of looking back, one may also realize that there are interesting alternative ways of approaching the problem. One may also come to appreciate that there is more to the problem than one first realized.

It is the latter observation that comes close to what it is that both Sowder **(The Looking Back Step in Problem Solving)** and Jones and Shaw **(Reopening the Equilateral Triangle Problem: What Happens If . . .)** have in mind when they focus upon *looking back* as a problem *posing* heuristic. Sowder takes a look at a square in which a diagonal and a median line are drawn. The intersecting lines create four regions and Sowder suggests some problem solving strategies (making use of symmetry, for example) that enable one to figure out the ratios of the areas of those four regions.

After coming up with such an analysis, however, Sowder suggests that in an effort to arrive at a better understanding of the nature of the solution, one should do a What-If-Not on the square as an attribute. He considers the cases of the rectangles, the rhombus and the parallelogram and comes up with some fascinating results.

It is worth keeping in mind, however that after modifying the initial attribute of squareness, he did ask the same question about the four regions of the new figures. In the spirit of What-If-Not, you might find it interesting to explore other questions as well. What are some of the questions you might ask after varying that attribute. As a start you might notice that the problem focuses upon *area*. That of course is a clue either that regions whose areas were previously unexplored might be investigated in all of the problems (including the original) or that something other than area (like perimeter?) might be investigated in all the cases.

You might also want to ask yourself what there is about the geometry of all four figures that accounts for the important commonality in the results. Is there something deeper going on? The suggestion that unexpected results might be a motivation for investigating *why* something is the case even after we supposedly have proven that it to be so is discussed in Chapter 6 of **The Art of Problem Posing.**

Jones and Shaw choose a different geometric example and engage in the same kind of activity. They look at the following problem:

> Given an equilateral triangle with any point P in the interior of the triangle, what is the sum of the distances from P to the three sides of the triangle?

If you have not looked at this problem before, you might want to explore it a bit before reading the article. It is something you could explore experimentally by actually measuring the lengths — especially if you have access to a program such as the Geometric Supposer (discussed in the article by Chazan in the next section of this chapter).

A clever construction and a little algebra reveals an answer that is a bit surprising. Motivated in part by this surprise, the authors set out to determine whether or not the location of P *inside* the equilateral triangle accounts for the unusual results.

They wonder if there might not be a kind of unity in the different alternatives they investigate and they speak of the frustration and floundering that engulfs them as they come close but have difficulty locating that unity.

The latter point is important pedagogically. What kind of opportunities do we provide for our students to become aware of and to express their frustrations? There might be important psychological and even mathematical advantages in encouraging our students to express their emotional reactions to all phases of inquiry. As a matter of fact, there are some emotions that are so closely tied to cognition that it would in effect destroy inquiry if they were non-existent. We have alluded to one such emotion in our discussion earlier—that of surprise (see Scheffler, 1977). Even the act of frustration is based upon some important awareness and is not merely an emotional phenomenon. We discussed some of these matters in the excerpt from **The Art of Problem Posing** that appears as the first piece in the Reflective section of this collection.

Jones and Shaw end their piece with some very interesting What-If-Notting on the "given" in their example. As in the previous case, however, they do maintain an the essence of the original question: What is the sum of certain segments from a given point connected with a given geometric figure? You might in fact inspire a whole new avenue of inquiry by changing the focus from sum to some other relation or operation. As in the previous case, it is of course worthwhile to maintain the original question—in the context of many variations of the phenomenon investigated—if one is primarily concerned with better understanding the original phenomenon.

In the third essay of this section, **Mathematics and Humanistic Themes: Sum Considerations,** Brown explores the famous problem attributed to Gauss: What is the sum of the natural numbers from 1 to n? As in the previous two essays, he is inspired to investigate the problem in greater depth by virtue of the fact that there are some unusual surprises as the analysis unfolds. That is, what appear to be conceptually different approaches arrive at the same result. The analysis for the sum in the case of n being odd for example eventually yields the same formula as the sum for the case of n being even—despite the fact that the unity seems to be a quirk of algebra rather than something that is inherent in the problem.

Brown asks *why* all of this works out as it does. Notice that the question is asked despite the fact that a possible answer might be: "It works because of the algebra." He comes to a better understanding of the problem and the unity of the disparate methods of solution by eventually engaging in the problem posing heuristic mentioned at the beginning of this chapter. That is, he asks the question: "How might we view this problem geometrically"?

In addition to raising such issues connected with *why,* this article is an early curriculum piece that focuses upon a topic that has grown in popularity in recent years: *a humanistic view of mathematics.* There is a newsletter entitled *The Humanistic Mathematics Network* (edited by Alvin White from Harvey Mudd College) that has been published since 1986. Though there are many interpretations of the underlying concept, Brown's article belongs to a tradition which raises questions regarding what it is we might learn about ourselves and our personhood in the context of doing and reflecting upon our mathematical behavior. Concepts such as *genealogy* and *potential* are introduced. The Brown essay not only focuses on "looking back" but on "looking inside oneself" in the context of "look back." It also intertwines a number of issues relating problem posing to problem solving, and is therefore a transition piece between this subsection and the one that follows it.

## REFERENCES

Polya, G. (1973). *How to solve it*. Princeton: Princeton University Press.
Polya, G. (1962). *Mathematical discovery: On understanding, learning and teaching problem solving*, Vol 1, New York: John Wiley & Sons.
Scheffler, I. (1977). In praise of the cognitive emotions. *Teachers College Record, 79,* 171–186.

# 24 The Looking-Back Step in Problem Solving

Larry Sowder

The prominence given to problem solving in *An Agenda for Action* (NCTM 1980) no doubt deserves much credit for the revitalized interest in problem solving in the curriculum. National and regional meetings of mathematics teachers have been featuring sessions on problem solving. Many elementary school text series are laying a problem-solving foundation with attention to a variety of heuristics. Soon we may no longer have to refer to the teaching of problem solving as our "lip-service objective" (Sowder 1972).

Many approaches to the solving of problems adapt, or build on, Polya's steps: understand the problem, devise a plan, carry out the plan, and look back (Polya 1957). Much teaching effort properly goes into developing several heuristics that may help in the devising-a-plan step. What may be neglected, however, is the last step, looking back.

Too much material always needs to be "covered," so it is easy to see why even a problem-solving advocate might devote inadequate time to the looking-back step. The solution of a problem is quite often followed only by "How about the next one?" Even such important questions as the following may be neglected: "Have we answered the question?" or "Is our answer reasonable?" We may too rarely ask, "How did we solve it?" to give emphasis and explicit attention to particular heuristics, or, "Is there another way?" to emphasize that problems may have more than one solution. It should be noted that Kantowski (1977) found that her ninth-grade geometry students did not spontaneously look back for alternate solutions and proofs, even after extensive instruction. She hypothesized that more experienced students might be more motivated to look for other solutions.

But looking back has even more to offer than just the possibility of finding a more elegant or simpler solution. Looking back can give our students a glimpse at an exciting part of mathematics, the creation of conjectures. Looking back can give our students a small taste of mathematics in the making, as opposed to the consumption of polished mathematics. Looking back can develop the outlook that how one gets answers is more important than the answers. Here is an example.

*Problem.*   Given square *ABCD*, with *M* the midpoint of side $\overline{AB}$. What are the ratios of the areas of the four regions defined by diagonal $\overline{AC}$ and $\overline{DM}$? (See fig. 24.1.) The reader may choose to solve the problem before reading the brief descriptions of several approaches in the next paragraph.

Solvers usually focus on regions I and III first, notice that they are similar, and then find the ratio of the areas of these two regions. Further progress follows different courses, with regions I and II a common next target. The theorem

> *the bisector of an angle of a triangle divides the opposite side into segments whose ratio equals the ratio of the segments defining the angle*

often appears in high school geometry texts and can be applied. (It usually is not thought of, unless it has appeared quite recently.) A coordinate approach seems to be more common. Even more elegant is noticing that $\overleftrightarrow{AC}$ is a line of symmetry for the square, which leads to the ratio for regions I and II. Once the ratio for regions I and II is found, the rest follows by noting that regions III and II make up half the square region, as do regions I and IV. Putting all these pieces together gives the answer, 1 : 2 : 4 : 5.

## Looking Back

The excellent articles and books by Brown and Walter (1970, 1983), by Walter and Brown (1969), and by Mason, Burton, and Stacey (1982) give many examples of the richness of pursuing mathematical tasks (or even situations) in a search for additional questions. Brown and Walter in

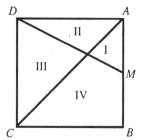

FIG. 24.1.

particular persuasively advocate the posing of problems through their "what- if-not" procedure, in which attributes of a situation are systematically varied to create new situations and questions, as a means of gaining greater insight into the situation and a greater appreciation for mathematical thinking. Such an approach to the problem here could lead, for example, to new problems about the perimeters of the regions (instead of the areas) or about cubes and cutting planes (instead of squares and lines).

A special case of the what-if-not procedure is "Can the problem be generalized?" A generalization of a result includes that result as a special case, so what-if-not explorations do not always yield generalizations. Inexperienced students, of course, could not be asked to generalize until after work with the what-if-not procedure or with questions like "Can you make up a problem similar to this one?" or "Does this make you think of another problem?"

Here are some generalizations of our problem that might be explored:

1. What would the ratios be if the figure were an $m \times n$ rectangle? (*Answer:* 1 : 2 : 4 : 5 also!)
2. What would the ratios be if the figure were a rhombus? (*Answer:* 1 : 2 : 4 : 5 again!)
3. What would the ratios be if the figure were an $m \times n$ parallelogram? (*Answer:* Still 1 : 2 : 4 : 5!)

An important lesson for the students is that the earlier ratios do not automatically apply to these (new) problems. Once articulated, which of these new problems one first pursues is probably dictated by one's method of solution of the original problem. For example, those who used coordinate methods on the square problem would find the first generalization attractive, since the right angles, which may have suggested a coordinate approach in the first place, are still there. Users of the angle-bisector theorem or the line of symmetry might find the second generalization more likely. The third conjecture, the most general of the three, admittedly appears less appealing, with the loss of both the right angles and the line of symmetry or angle bisector.

It is probably surprising that the same ratios hold for the new problems 1, 2, and 3, independent of the dimensions of the figure. That fact in itself invites looking back; can we now see (intuitively) why the ratios are the same? The areas, first thought of in terms of squares, could be expressed in terms of the rhombi, rectangles, or parallelograms that would result by tilting or stretching the original square (and the embedded square units).

These new problems all resulted from generalizing on the given figure. One could also vary point $M$:

4. What would the ratios be if $M$ were a point of trisection with, for example, $AB = 3AM$? (*Answer:* $1 : 3 : 9 : 11$)
5. What would the ratios be if $M$ gave segments in the ratio $r : (1 - r)$, with $0 < r < 1$? (*Answer:* $1 : x : x^2 : (x^2 + x - 1)$, where $x = 1/r$)

An even more ambitious conjecture could change both the figure and the point of division:

6. What would the ratios be if the figure was a parallelogram and $M$ a point of arbitrary division? (*Answer:* The same as in problem 5)

What is perhaps most exciting is finding that one or more of these new problems yields to the same method of solution that one has used earlier. Indeed, one might claim that the "best" method for solving a problem is the one that carries over to generalizations of the problem. If the solutions to the new problems also turn out to be surprising, as happens here, so much the better!

## SUMMARY

Looking back at a problem after its solution should at least on occasion include the generation of new problems. The students then are making up mathematics, rather than merely absorbing it, and are being exposed to an important and exciting part of mathematical thinking—the generation of new ideas. Another problem useful for generating new ideas is the "cookie cutter" problem: What is the greatest number of regions determined by five congruent circles in a plane? (Bright 1977; Schwartzman 1977)

## REFERENCES

Bright, G. W. (1977). Reader reflections: Learning to count in geometry. *Mathematics Teacher, 70,* 15–17.

Brown, S. I., & Walter, M. I. (1983). *The art of problem posing.* Philadelphia: The Franklin Institute Press.

Brown, S. I., & Walter, M. I. (1970). What if not? An elaboration and second illustration. *Mathematics Teaching, 51,* 9–17.

Kantowski, M. G. (1977). Processes involved in mathematical problem solving. *Journal for Research in Mathematics Education, 8,* 163–180.

Mason, J., Burton, L., Stacey, K. (1982). *Thinking Mathematically,* London: Addison-Wesley.

National Council of Teachers of Mathematics. (1980). *An Agenda for Action: Recommendations of the 1980s.* Reston, VA: Author.

Polya, G. (1957). *How to solve it.* (2nd Ed.). Garden City, NY: Doubleday.

Schwartzman, S. (1977). Reader reflections: One sided polygon. *Mathematics Teacher, 70,* 564.

Sowder, L. (1972). Teaching problem-solving: Our lip-service objective. *School Science and Mathematics, 72,* 113–116.

Walter, M. I., & Brown, S. I. (1969). What if not? *Mathematics Teaching. 46,* 38–45.

# 25 Reopening the Equilateral Triangle Problem: What Happens If . . .

Douglas L. Jones
Kenneth L. Shaw

An interesting, instructive, and classic problem that offers a rich setting for exploration is the equilateral-triangle problem:

> Given an equilateral $\triangle ABC$ and a point $P$ inside $\triangle ABC$, what is the sum of the distances from $P$ to the three sides?

If we refer to figure 25.1, the problem can be restated as follows:

> "Find $PD + PE + PF$."

This problem offers a wonderful setting in which to employ some of the heuristics that Pólya advocated (1973, 108–9, 114), namely, try a simpler or related problem and then generalize. Students who use these techniques

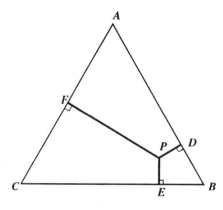

FIG. 25.1.  Find $PD + PE + PF$.

often convince themselves that $PD + PE + PF = h$, where $h$ is the altitude of the triangle. Further exploration may lead to a proof of this conjecture based on summing areas of the partitioned triangle as seen in figure 25.2.

area $\triangle PAB$ + area $\triangle PBC$ + area $\triangle PAC$ = area $\triangle ABC$

If $s$ refers to the length of a side of $\triangle ABC$, we have

$$\frac{s(PD)}{2} + \frac{s(PE)}{2} + \frac{s(PF)}{2} = \frac{s(h)}{2},$$

$$PD + PE + PF = h.$$

Since this formula requires only that $P$ be located within $\triangle ABC$ and not in any certain position, the problem is solved. Or is it?

The importance of problem solving has been an emphasis in mathematics education for many decades, but perhaps never so strongly as since *An Agenda for Action* (NCTM 1980, 2–5) proclaimed problem solving to be the centerpiece for teaching mathematics in the 1980s. Much has been written about the importance of problem solving (e.g., Krulik and Reys [1980]; Dessart and Suydam [1983]; Pólya [1962, 1973]) and how problem solving can be used in the classroom (e.g., Halmos [1975]; Lesh [1981]; Schoenfeld [1982]). Brown and Walter (1983) argue that problem solving should focus not only on how a problem is solved but also on the questions asked prior to and while working on the problem. The use of the question, "What happens if . . .?" is one strategy that can help students see important aspects of a problem that they may otherwise have overlooked. The remainder of this article focuses on the use of the question, "What happens if . . .?" as a means of problem posing and as a way students can participate in the creation of their own mathematical ideas.

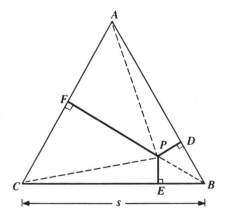

FIG. 25.2.   Area $\triangle PAB$ + area $\triangle PBC$ + area $\triangle PAC$ = area $\triangle ABC$.

What happens if, in the equilateral-triangle problem, the point $P$ is *not* located inside $\triangle ABC$?

This possibility places the problem in a dynamic setting and motivates us to explore it further. What follows are investigations that can be made into the different options suggested by the previous question.

## OPTION 1

What happens if $P$ is *on* $\triangle ABC$? (See fig. 25.3.)

## OPTION 2

What happens if $P$ is in the exterior of $\triangle ABC$?

*Case 1.* $P$ could lie in the interior of an angle of $\triangle ABC$, but not inside $\triangle ABC$. See $P_1$ in figure 25.4.

*Case 2.* $P$ could lie in the interior of a vertical angle associated with one of the angles of $\triangle ABC$. See $P_2$ in figure 25.4.

Would it still be true that the sum of the lengths of the perpendiculars from $P$ to the sides (or their extensions) is the length of the altitude of $\triangle ABC$? By looking at figure 25.5 and other examples, it is clear that it would not. However, we believe some linear combination of these lengths would yield the length of the altitude, that is,

$$a(PD) + b(PE) + c(PF) = h$$

for some real numbers $a$, $b$, and $c$.

*Exploration of option 1.* Let $P$ be on $\triangle ABC$ as seen in figure 25.3. With this option only two perpendiculars exist, $\overline{PD}$ and $\overline{PF}$. A geometric proof

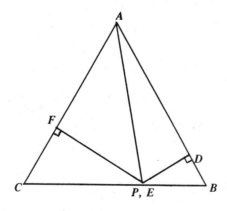

FIG. 25.3.    *P* on $\triangle ABC$

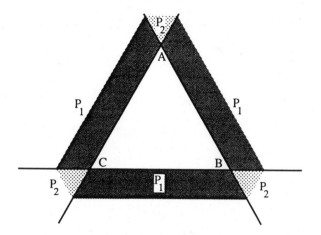

FIG. 25.4.   *P* in the exterior of △*ABC*

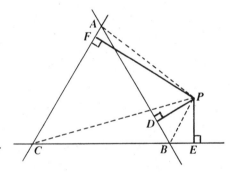

FIG. 25.5.   *P* lies in the interior
of an angle of △*ABC*.

similar to that used in the original problem is used to show that *PD* + *PF* = *h*.

area △*PAB* + area △*PCA* = area △*ABC*

$$\frac{s(PD)}{2} + \frac{s(PF)}{2} = \frac{s(h)}{2}$$

*PD* + *PF* = *h*

In some sense, this option is a limiting case of the original equilateral-triangle problem. Since perpendicular $\overline{PE}$ has length 0, it effectively reduces the number of perpendiculars from three to two. In addition, it yields a linear combination of the lengths of the perpendiculars that is equal to the length of the altitude. A still more limiting case occurs when the point *P* lies on a vertex of △*ABC*. In this case, two of the perpendiculars have length 0; the other is an altitude of the triangle.

*Exploration of option 2, case 1.*   Let $P$ lie in the interior of an angle of $\triangle ABC$, but not inside $\triangle ABC$. See figure 25.5.

We construct the perpendiculars from $P$ to the sides of the triangle or their extensions and draw a dotted segment between $P$ and each vertex. We note the following:

area $\triangle PAC$ + area $\triangle PCB$ − area $\triangle PAB$ = area $\triangle ABC$

$$\frac{s(PF)}{2} + \frac{s(PE)}{2} - \frac{s(PD)}{2} = \frac{s(h)}{2},$$

$$PF + PE - PD = h$$

This formula represents a linear combination of the lengths of the perpendiculars that equals the length of the altitude. By this time, we are wondering if the linear combination might even be so special that the coefficients $a$, $b$, and $c$ would be either 1 or −1. We are concerned that the location of $P$ will determine which of the lengths will be positive and which will be negative. We conjecture that the determining factor is whether the foot of the perpendicular intersects a side of the triangle or whether it intersects the extension of that side. The diagrams in figures 25.6 and 25.7 show these different regions; the numbers indicate how many sides of $\triangle ABC$ the feet of the perpendiculars from $P$ will intersect if $P$ is in that region. For example, in figure 25.6 they intersect one side and the extensions of the other two sides, but in figure 25.7 they intersect two sides and the extension of the third side. Our conjecture was easily shown to be

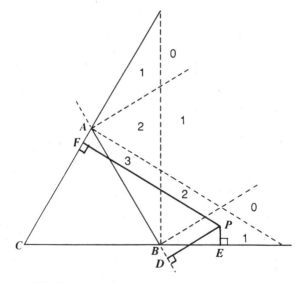

FIG. 25.6.   Intersecting one side and two extensions

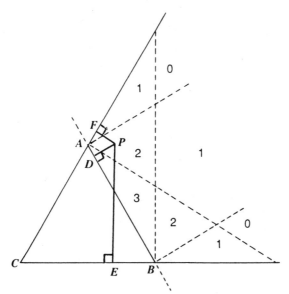

FIG. 25.7.    Intersecting two sides and one extension

invalid as seen in figure 25.5. Notice that *PF* and *PE* have coefficients of 1 in *PF* + *PE* − *PD* = *h* but that the foot of $\overline{PF}$ intersects a side, whereas the foot of $\overline{PE}$ intersects an extension. Even though we are still unsure of how the location of *P* affects the coefficients of *PF*, *PE*, and *PD*, we hope to gain more insight by exploring case 2.

*Exploration of option 2, case 2.*   Let *P* lie in the interior of a vertical angle associated with one of the angles of △*ABC*. See figure 25.8.
We again construct the perpendiculars to the sides of △*ABC* or their

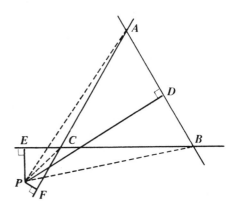

FIG. 25.8.    *P* lies in the interior of vertical ∠ *ECF* with respect to ∠ *ACB*.

extensions and draw the dotted segments between $P$ and each vertex. This process results in the following equations:

$$\text{area } \triangle PAB - \text{area } \triangle PAC - \text{area } \triangle PCB = \text{area } \triangle ABC$$

$$\frac{s(PD)}{2} - \frac{s(PE)}{2} - \frac{s(PF)}{2} = \frac{s(h)}{2},$$

$$PD - PF - PE = h$$

A third case in which $P$ lies on the extension of one of the sides is a limiting case for this option. This investigation is left for the reader.

This linear combination seems to affirm in our minds the fact that the location of $P$ does determine which of the lengths will be treated as positive and which will be treated as negative. We are convinced that we can always find a linear combination of the lengths of the perpendiculars that will equal the length of the triangle's altitude, but we are bothered by the fact that it seems to depend on first locating the point $P$. We feel we must be able to make a generalization that will replace these special cases and satisfy the original problem.

At this point, we are unable to see a way to a generalization. In effect, we are floundering. Our consideration of the various cases and options has been an attempt to superimpose some structure on the problem. Although we feel that the floundering is a valuable part of our investigation, we are frustrated with our seeming lack of progress; we are not able to find a pattern in which lengths will be positive and in which they will be negative.

By looking at the figures we have drawn and thinking about the various cases we have investigated, it becomes clear that the sign of a distance in the linear combination is based on the relative position of the side to which its perpendicular is drawn and the vertex opposite that side. Our generalization, illustrated in figures 25.9–25.11, is given as follows.

Let $\triangle V_1 V_2 V_3$ be an equilateral triangle with altitude $h$. Denote by $s_i$ the side opposite vertex $V_i$, where $i$ ranges from 1 to 3. If $P$ is any point in the

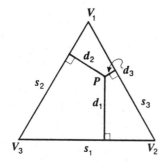

FIG. 25.9. Example 1 (original problem): $1\, d_1 + 1\, d_2 + 1\, d_3 = h$.

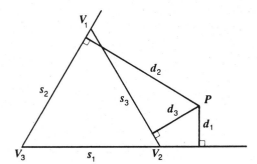

FIG. 25.10.   Example 2 (option 2, case 1): $1\,d_1 + 1\,d_2 - 1\,d_3 = h$.

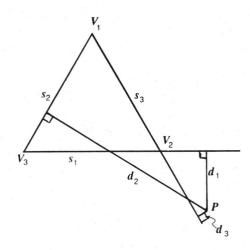

FIG. 25.11.   Example 3 (option 2, case 2): $-1\,d_1 + 1\,d_2 - 1\,d_3 = h$.

plane of $\triangle V_1 V_2 V_3$ with $d_i$ equal to the perpendicular distance from $P$ to $s_i$ or its extension, then

$$k_1 d_1 + k_2 d_2 + k_3 d_3 = h,$$

where $k_i = \begin{cases} -1 \text{ if } P \text{ and } V_1 \text{ are on opposite sides of the line containing } s_1 \\ 1 \text{ if } P \text{ and } V_1 \text{ are on the same side of the line containing } s_1 \text{ or} \\ \text{if } P \text{ is on } s_1. \end{cases}$

A proof of this proposition based on summing areas is straightforward and is left for the reader.

This article has demonstrated that mathematical problems can furnish rich settings for investigation and need not be used only for seeking particular solutions. Pólya (1973, 14–19) has suggested that a critical phase of problem solving is looking back at the solution. He not only stressed the need for checking the solution but also emphasized applying both the result and the method to other problems. In looking back at the equilateral triangle problem, we can see other problems to which this result or method could be applied. The following are some examples:

1. What happens if $\triangle ABC$ is not equilateral?
2. What happens if the problem is extended to any regular $n$-gon?
3. What happens if the problem is not restricted to the plane? For example, can a general solution be found for a tetrahedron?
4. What happens if the distances involved are from $P$ to the vertices rather than from $P$ to the sides?
5. What happens if the segments $\overline{PD}$, $\overline{PE}$, and $\overline{PF}$ are not perpendicular to their respective sides but rather make an angle of 50°?

In a sense, the equilateral triangle problem is merely a placeholder. We have used it to illustrate the generative nature of problem posing using the heuristic "What happens if . . .?" We feel that extending a problem beyond its given constraints and context adds a dynamic aspect and suggests many new insights into the original problem. Within this exploratory framework, the understanding of the mathematics in which the problem is embedded deepens.

## REFERENCES

Brown, S. I., & Walter, M. I. (1983). *The art of problem posing.* Philadelphia: The Franklin Institute Press.
Dessart, D., & Suydam, M. (1983). Problem solving. In *Classroom ideas from research on secondary school mathematics* (pp. 29–38). Reston, VA: National Council of Teachers of Mathematics.
Halmos, P. (1975). The teaching of problem solving. *American Mathematical Monthly, 82,* 446–470.
Krulik, S., & Reys, R. E. (Ed.). (1980). *Problem solving in school mathematics: 1980 Yearbook of the national council of Teachers of Mathematics.* Reston, VA: National Council of Teachers of Mathematics.
Lesh, R. (1981). Applied mathematical problem solving. *Educational Studies in Mathematics, 12,* 235–64.
National Council of Teachers of Mathematics. (1980). *An Agenda for Action: Recommendations of the 1980s.* Reston, VA: Author.
Polya, G. (1973). *How to solve it.* Princeton: Princeton University Press.
Polya, G. (1962). *Mathematical discovery: On understanding, learning and teaching problem solving* (Vol 1, 2). New York: John Wiley & Sons.
Schoenfeld, A. (1982). Measures of problem-solving performance and of problem-solving instruction. *Journal for Research in mathematics Education, 13,* 31–49.

# 26 Mathematics and Humanistic Themes: Sum Considerations

Stephen I. Brown

## I. INTRODUCTION

If one starts with the hypothesis that aspects of mathematics learning could have an impact on the way one conducts one's life and views the world in other domains as well, what are some of the possible mind-expanding ways in which it might be conceived? It is perhaps long overdue for educators to ask themselves some daring questions with regard to possible humanistic import of the disciplines and in particular with regard to mathematics. It is unquestionable that the "structure of the disciplines" movement of the past fifteen years solved some problems in the past decade for some people. Especially in mathematics one was able to arrange the discipline so that its alleged essence as an axiomatic subject was apparent for many students. Supposedly there was de-emphasis of such bugaboos as rote memory and drill and a renaissance of understanding and discovery (see Brown (1971) for a discussion and criticism of these issues).

No matter how exciting and successful such an enterprise might have been — and the matter is quite controversial — this development of the past decade and a half was bound to be uneventful from at least one point of view: impact on students as human beings living in a complicated and problematic world. The basic question asked by curriculum designers was: How might I best convey the nature of mathematics (especially the twentieth century spirit) to youngsters? A much more daring question appears to us to be: How might we use mathematics (among an arsenal of other things) to convey knowledge and attitudes towards the world and about oneself that would be valuable in many non-mathematical contexts?

We are neither advocating a return to faculty psychology; nor do we mean to focus on "applications of mathematics to the real world." We are trying to get at something much deeper. One point of extreme irony has been that the more we focus on mathematics as a tool for dealing with the world in humanistic terms, the closer we seem to come to a kind of training that if appropriately applied might also produce first rate mathematicians.

That there are many potentially negative psychological or epistemological attitudes conveyed to students at all levels even by "modern mathematics" programs is obvious to anyone who would bother to look. Instead of discussing them at this point, we shall refer to them in the context of presenting what we believe are healthy alternatives. By choosing a small number of elementary mathematical concepts we hope not only to suggest alternatives, but to consider a number of worthwhile research issues, and to point towards philosophical problems as well.

As we move through the paper we will be shifting from problems related to mathematics as a public "out there" phenomenon to issues that enable us to view our relationship to the discipline in more personal terms. We are not suggesting that these are the only issues that may be raised. We only wish to illustrate how one might begin to think of the use of mathematics as a humanistic enterprise. We shall begin the "out there" analysis in Section II by focusing on one rather elementary problem to illustrate the potential of mathematics: finding the sum of all integers from 1 to n. We intend to show how one might profit from an examination of the sum problem so as to exemplify a spirit of inquiry that we feel is one very important and one very neglected aspect of intelligent and creative thought. We beg the reader to participate first hand in the details of Section II before searching for the significance which we discuss at the end of that section.

## II. FORMAL, SURFACE AND DEEP CONNECTIONS[1]

### One Problem:

What is the sum of the following series?

$$1 + 2 + 3 + \ldots + 98 + 99 + 100$$

The reader will perhaps recognize this as the famous problem allegedly solved by the great mathematician Gauss in a few seconds after his fifth grade teacher assigned it as "busy work." Wertheimer (1945) has spoken at length about this problem and has used it to explain notions related to Gestalt psychology. In particular, he conveys the distinction between a

---

[1]With regard to a different problem some of the ideas of this section appear in Brown & Keren (1972).

"good" and a "bad" gestalt. A good gestalt, allows the person to make observations such as the following:

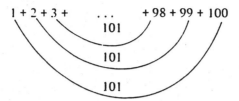

Instead of unrelated entities, we find that we have a constant here that is significant: namely a certain number of pairs of sums (50) that are each the same (101). So the solution is clearly (50) × (101).

He, however, goes beyond the distinction between "good" and "bad" gestalt in order to make another worthwhile psychological point as he distinguishes between logical and psychological equivalence. Consider two solutions to the following modified problem:

What is 1 + 2 + 3 + 4 + 5 + 6?

**Two Different Solutions:**

As with 100 terms, a solution could obviously be derived as follows:

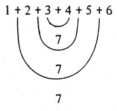

Therefore the sum is: (6 + 1) × (3), or generalized to n terms, it would be:

$$(n + 1) \bullet \frac{n}{2}$$

Let us call this the "pure pair" solution. We could, as Wertheimer points out, view the problem slightly differently. We could claim that there really is a "fulcrum" in this problem; that 3.5 is a center. Here we have a "teeter totter" in which every term to the left of the fulcrum and at a given distance from the fulcrum is less than 3.5 by the same amount that every term to the right and at the same distance exceeds 3.5. The following picture is suggested:

1 + 2 + 3 +   + 4 + 5 + 6

As the picture suggests, since whatever exceeds 3.5 on one side is compensated for on the other, we might as well consider each term to be 3.5. Therefore the series might as well look like:

$$3.5 + 3.5 + 3.5 + 3.5 + 3.5 + 3.5$$

So: the sum is obviously: (3.5) × (6), or generalized to n terms it would be:

$$\frac{(n + 1)}{2} \bullet n$$

Let us call this the "repeated imagined middle term" solution. Let us look now at the two general formulas for the "pure pair" and the "repeated imagined middle term." It is clear that they must be logically equivalent since both of them answer the question:

$$1 + 2 + 3 + \ldots + (n - 1) + n = ?,$$

and the sum is always the same for a fixed n. Furthermore it is obvious that they are equivalent, for we can prove algebraically that

$$\left(\frac{a}{b}\right) \bullet c = a \bullet \left(\frac{c}{b}\right)$$

the forms of the two general solutions. However, as Wertheimer points out, these general solutions are different psychologically since they are derived from different *perspectives*. Because of the relatively trivial nature involved in demonstrating the equality of the two algebraic expressions, we are not surprised to find that the "repeated imagined middle term" and the "pure pair" are logically the same. As we shall soon see, however, there are other ways of looking at the problem that would lead us to be surprised (and feel the need to verify the alleged equality we arrive at) despite the fact that these different expressions both exemplify the same phenomenon.

## A Modified Problem: The Two Previous Solutions Revisited:

Let us find the sum of a slightly different sequence:

$$1 + 2 + 3 + 4 + 5.$$

Once we have solved the previous problem in the two ways suggested, it should be child's play to do the above. Let us look at the problem in a method analogous to the "pure pair" solution:

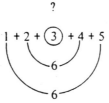

Here we have the same analysis in the case of the sum of 1 through 6 except for the disturbing intrusion of a *middle term* without a mate. But this problem is not as insurmountable as it appears on first glance. The middle term is merely *half* the sum of each of the other pairs,

$$\text{i.e. } 3 = \frac{5 + 1}{2} = \frac{4 + 2}{2}.$$

So, if we were to generalize for n terms, we would conclude that its formula should be:

$$(n + 1) \times \frac{(n - 1)}{2} + \frac{(n + 1)}{2}$$

|                                |                                    |                       |
| ------------------------------ | ---------------------------------- | --------------------- |
| the sum of<br>each pair        | the<br>number<br>of pairs          | the middle<br>term    |

Let us call this the "modified pure pair" method. Suppose we look at this problem through a method analogous to the "repeated imagined middle term" method. Then what is the fulcrum? Here it is even clearer what the fulcrum is than in the first problem. Instead of considering the middle term to be a "nuisance" because it does not have a mate, we use that fact as a starting point. 3 is in fact our real fulcrum, and we have

$$1 + 2 + 3 + 4 + 5$$

As with the analysis of the sum of 1 through 6, we see that for every term that exceeds 3 to the right, there is one to the left that compensates for it by the same amount, and so the above sum might as well be:

$$3 + 3 + 3 + 3 + 3.$$

The sum then would be (3) × (5) and if we generalize it for n terms, we would have:

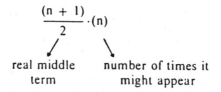

Let us call this the "repeated real middle term" solution.

## The Four Solutions Re-Examined:

Now let us re-examine the sum to 5 and the sum to 6 from several points of view. It was obvious on two counts why it is that the two generalizations of the solutions for the sum to 6 were the same: First of all they are both generalizations for the same problem (with n being the same value in each case), and secondly, it is obvious at a glance that

$$(n + 1) \cdot \frac{n}{2} = \frac{(n + 1)}{2} \cdot n.$$

But what of the case of the two solutions for the sum to 5? Again in both of the analyses, the n is the same, and since the sum of n terms is constant, we would expect them to be equivalent. But are they? Here we are not so certain. A little manipulation of the "modified pure pair" solution will reveal that the "repeated real middle term" and the "modified pure pair" solutions are indeed equivalent. Though we are relieved to discover the equality, we were perhaps more doubtful than in the case of the two equivalent expressions for the sum to 6.

Let us go one step further, however. Compare the four formulas we get for the sum to 6 and the sum to 5 problems. What's the major difference between these two problems? It is obvious that in one case we are generalizing from an even number of terms, and in the other from an odd number. Furthermore, though there are some similarities in the analysis between the two problems, it is also obvious that there are differences. For example, in the "modified pure pair" approach, for the first time we *tag* along an extra term, (n + 1)/2 because the middle one has no mate. We never did this in the analysis of the sum to 6 problem. Though we are not too surprised to discover that for the two different approaches for each case the formulas are equivalent, it is indeed a surprise to find out that the "pure pair" and the "modified pure pair" are equivalent. It is less of a shock of course to see, for example, that the "repeated imagined middle term" and the "repeated real middle term" are equivalent, because as a matter of fact from the point of view of *form,* they are identical. There is of course a difference in interpretation, for in the case of one we construct an "imagined" middle term, while for the other we deal with an actual middle term. So though we are not as shocked by the equivalence of these two forms (in the sense that we do not feel the urge to *prove* the equality) as we were for the first "cross breeding," we should be just as curious to know *why* this happens. Furthermore there is a sense in which no amount of proving algebraic equalities will help us answer the *why* of that question. Something more profound is needed. We must ask: Why should we expect the formula for the sum of an odd number of terms to be equivalent to that for an even number of terms?

### The Search for A Deep Connection Between Odd and Even n:

We are really asking here: Is there a *deep connection* between the posing of the problem for an odd number of terms and for an even number of terms? Before trying to analyze this question, let us observe the connections we have discovered. First of all, there is a *surface similarity* in both cases. That is, they both deal with sums of ascending consecutive numbers. These words themselves suggest that there might be a *deeper similarity.* Furthermore, though the analyses are *different,* the algebraic *form* of the two analyses is the same, as the two "cross breeding" experiments showed. Is there some way that we might see the odd and the even case as the same in a *deep* sense? What we are essentially searching for is an analysis that is independent of "parity" (oddness and evenness). Can we find one? We shall examine a deep analysis, and then return to a discussion in more general terms of the broad educational significance of several issues beneath the surface of this problem.

Anyone who has taken mathematics through intermediate algebra most likely has proved the more general formula for the sum of an arithmetic progression; that is, re-express

$$s = a + (a + d) + (a + 2d) + \ldots + (a + (n - 1)d)$$

for an arithmetic series of n terms. The following "device" is then used. We write the series backwards underneath the forward series:

$$s = a \qquad\quad + (a + d) \qquad + \ldots + (a + (n - 2)d) + (a + (n - 1)d)$$
$$(a + (n - 1)d) + (a + (n - 2)d) + \ldots + (a + d) \qquad\quad + a$$

Then,

$$2s = n(2a + (n - 1)d) \quad .$$

and

$$s = \frac{n(2a + (n - 1)d)}{2}$$

Applying this technique to the special case we are dealing with, we would analyze it as follows:

$$s = 1 + 2 + \ldots + (n - 1) + n$$
$$s = n + (n - 1) + \ldots 2\quad + 1$$

$$2s = (1 + n)n$$

$$s = \frac{n(n + 1)}{2}$$

In one sense this is the deep connection we have been looking for. It is obvious that the above formulation does *not* depend upon parity for the argument in that it is the same for the case of odd as for even.

## A Possible Motivation for The Deep Proof:

On the other hand, something is missing from this analysis as Wertheimer points out very well. Why would one ever have been tempted to write the sum in two different ways as above? What's behind the algebra which is obviously correct? What could have been the incentive for such a scheme? A look at the *form* of the final formula,

$$\frac{n \bullet (n + 1)}{2}$$

might suggest how we could conceive of the problem in terms that would be more meaningful for many people. What does the form remind one of? There are many things it might bring to mind, but one that turns out to be particularly helpful is the formula for the area of a triangle. How could we connect up that formula with the area of a triangle in a meaningful way? If we think of each number as represented by that number of congruent squares, then there is one very interesting interpretation. The diagram below suggests one possibility.

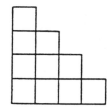

Here we essentially have a right triangle. But how should we represent the sum in a meaningful way? What is the significance of writing the sum *twice*

in the algebraic proof? Some degree of manipulation with squares together with the insight that

$$\frac{n \bullet (n + 1)}{2}$$

has its form to be that of the area of a triangle might suggest the following:

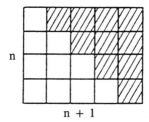

That is, if we take a congruent copy of the original figure and place it on top and to the right of the given figure, we get a rectangle. Obviously since the entire figure is a rectangle, its area is $n \bullet (n + 1)$. Since we only want the area of the triangle, and since the two figures making up the rectangle are congruent, its area is

$$\frac{n \bullet (n + 1)}{2}$$

**Additional Spin-off of the Geometric Motivation:**

Having seen one proof motivated by a geometric staircase model, we might search for modifications of that model which yield different (but of course logically equivalent) algebraic formulations. Two such models (which do not depend upon parity) are sketched below.

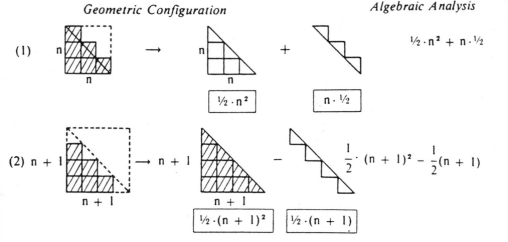

Before turning to educational implications we should mention that none of the algebraic analysis depends upon (though of course they can be used unnecessarily) sophisticated notions of area and accompanying formulas. The notions that are needed are of the following order: Congruent figures have the same area; the diagonal divides a rectangle into congruent halves. The only formula we use explicitly in the algebraic analysis is that the area of a square whose side is n is $n^2$.

## Educational Implications:

What can we learn from this problem? Wertheimer and other gestalt psychologists in analysing this problem have placed primary emphasis upon blind and deep discoveries and applaud approaches that are insightful. Thus Wertheimer finds fault with students who approach the problem of finding a desired sum by combining two sums written beneath each other and handled algebraically. He would claim that though technically correct, it is blind to inner meanings of the ideas.

I think his emphasis is misguided, though he provides an intelligent first approximation for the analysis of an important educational issue. Notice that the *form* of the deep *answer* gotten by writing the algebraic sum twice was what might have inspired us to search for a geometric solution that is *deep* in the sense that it does not depend upon parity.

It is *not* that we ought to abhor and eliminate solutions that lack *depth* in the sense that they do not get at the inner workings of the mathematics. Instead we would like to suggest as a worthwhile educational program the explicit analysis of problems along the dimensions of *surface, deep* and *formal* similarity and differences. When is the similarity in approach only one of *surface* similarity (e.g. when the only connection might be that the same words are used for the problems that otherwise share no common ground)? When is the connection a *formal* one? When *deep*? More importantly, how can we learn to encourage valuable thought by searching for *interaction* among and between these three categories? A brief recapitulation of our line of attack for the sum problem should indicate the potential of such interaction.

Recall that in the two different approaches to the odd problem there is a *deep* connection in that the symbols stand for the same idea, and thus they must be logically equivalent. The same is true in the two approaches to the even problem. There is, however, no *deep* "cross breeding" connection between the even and odd approaches. Since the *forms* are very similar (as a matter of fact, logically equivalent) we were inspired, however, to search for a deep connection. Thus a connection which is similar only in *form* or *surface* can inspire a search for *deep* similarity. Furthermore a *deep* connection (the geometric double sum approach) can be examined and modified in order to find other possible interpretations for *forms* (algebraic) that may already be invested with meaning (which we have not done) as well as to encourage a search for new forms (which we did in the preceding sub-section).

We have in this section chosen one example to illustrate a heuristic for problem solving and not to demonstrate the best way (in some sense) of solving the problem. Thinking of this as an "ideal" case for the purposes of such illustration, we have not indicated all the side paths and erroneous conceptions that in fact occur when real people solve this problem. How fruitful are the three categories of comparison? What are their logical and psychological limitations?[2] To what extent do the comparisons of similarities and differences along the lines we have suggested require a self-conscious attitude, and to what extent can they be left unverbalized? Though comparison is obviously not the only thing one does in creative thought, it is at least one element. How fairly this element has been characterized is a topic the author welcomes criticism on. In what way might interdisciplinary curricula be devised around this particular focus? That is, is the notion of comparison of similarities and differences (especially along the three dimensions I have suggested) as central as I have been suggesting for creative and critical thought in general?

## III. GENEALOGY AND POTENTIAL

### A Student's Observations on Elegance:

After discussing the various ideas in Section II with my class, a student commented that though discussion of forms was all well and good, we might certainly agree that the "non-insightful" algebraic approach (perhaps also conceding the geometric counterpart) was the most elegant, since it yielded a result that was both *general* (covering odd and even cases) and *compact* in the sense that it yielded a form that was easiest to remember for purposes of future application. It seems to me that the notion of elegance

---

[2]See Goodman (1972), especially Chapter IX 2, for an excellent discussion of philosophical problems related to the notion of similarity. In particular he has shown why it is that one cannot equate similarity with possession of common characteristics, but that in fact such characteristics must be weighted for importance (a context laden term). Worse yet, he has shown that the concept of *similarity with regard to characteristic X* (regardless of importance) does not so much modify similarity and further clarify the context as it makes the notion of similarity redundant. If in fact the notion of similarity is as philosophically limp as Goodman claims, that does not automatically damage the educational claims I have made. It at least suggests however, that we might wish to be more careful about what it is we are claiming of a psychological or educational nature. We shall leave that as a task for the future.

runs much deeper than his intuitive conceptions, and I should like to explore a number of aspects of the concept in order to get some idea of its dimensions. We are not out to come up with a formal definition, but in establishing a prolegomenon (inaccurate as it may be) I hope that we shall generate humanistic issues that themselves go considerably beyond the notion of elegance.

## A First Approximation:

An elementary example that comes to many mathematicians' minds of a proof that is truly elegant is Euclid's proof of the infinitude of primes. Recall that a natural number n is prime if it has exactly two different divisors. Thus 2 is prime for only 1 and 2 divide it in the set of natural numbers. 3, 5, 7 are prime. 9 is not since there are *three* factors of it — 1, 3, 9.[3] Euclid was supposedly the first to suggest the following proof:

For any finite number of primes $P_1$, $P_2$, . . ., $P_n$ create a new number N as follows:

$$N = P_1 \cdot P_2 \cdot P_3 \cdot \ldots \cdot P_n + 1.$$

Either N is prime or composite. If it is prime, we have created a new larger number than any of the $P_i$'s that is prime. If it is composite, it must be divisible by some prime number (and this requires analysis but is not too hard to see intuitively). Now, none of the $P_i$'s so far listed will do, since they leave a remainder of 1 in each case. Hence there must exist at least one more prime.

What is elegant about this solution? It has a number of characteristics that might appeal to one searching for elegance:

1. It is brief.
2. It involves a creation that is unexpected given the machinery.
3. Once created, it is easy to understand and complete the proof.

One of course might want to add to the list for this example, and could of course come up with examples in which other criteria are introduced. In addition, it is clear that each of these criteria must be stated relative to the expertise and experience of the assessor.

Furthermore we ought to observe that though these may be neither necessary nor sufficient conditions for us to claim that a solution be an

---

[3]For some idea of the mathematical significance of the concept of primality as well as an indication of the sensitivity to domain of answers to questions about primes, see Brown (1965). For a more sophisticated mathematical analysis which also focuses on pedagogical questions, see Brown (1968).

elegant one, we would seem to require a conjunction of some such set of simple criteria (perhaps appropriately weighted) in order to make a legitimate elegance claim. Any one of the criteria in itself would obviously *not* do. We know many solutions that satisfy these criteria disjunctively that we would all agree are not elegant.

## A Second Look:

It should be obvious in the example of the infinitude of primes that we have been referring to elegance with regard to *one* aspect of mathematics, and the one that is most widely associated with mathematics — namely *proof.* In the case of proof, a solution which is elegant would of course have also to be a solution, and in that case, we have a truth component implied in (3).

Where else might we look for elegance in mathematics? We might look merely at the *statements* of theorems (or conjectures for that matter) in mathematics. What are some assertions that are elegant ones? One that comes to mind is the Fundamental Theorem of Integral Calculus. Here we are told that in order to evaluate a particular function — the definite integral we must consider which family of functions that function is a derivative of. That is, two notions that are conceptually distinct on the surface — the definite integral, intuitively seen as an area, and the derivative, seen as the slope of a tangent line to a curve at a point — are found to be linked in an unsuspected way. Analogous to (2) for *proof,* this *statement* connects up concepts that at first glance could not look more unrelated. Whether or not this characteristic is a necessary and sufficient one for elegance of *statements* in mathematics we leave for future analysis. Let us turn now to territory that is even less easily analyzed but is in many ways more interesting.

## Genealogy:

My student's claim that the algebraic solution was the most elegant rested in part I believe upon a notion of simplicity. Especially within the context of traditional mathematics, we were always to represent our result in *simplest* forms. Granted that an analysis of simplicity is "sticky," let us accept an inaccurate and oversimplified notion such as that which involves the fewest number of operations and/or variables. Once we are sensitized to quantification and logical equivalence we are less prone to place a simpler expression in higher esteem than a more complicated one. With this rather neutral stance with regard to expressions, however, we tend to overlook an important psychological feature that we alluded to in Section II. That is, to what extent does the final expression capture the ways in which we either thought about the problem or derived the results? Some of the rather messy expressions of Section II might be excellent reminders of how we conceive of the task or what we focused on in order to come up with the conclusions we did. Here in fact we might have an aspect of elegance that we have not previously considered. That is, in addition to looking at elegance of (1) *proof* or of (2) *statement alone* we might search for some aspects of *interaction* between the two. I am suggesting the notion of *genealogy* as one important dimension of elegance that we tend to ignore when we pass judgment on different approaches to mathematical problems. That is, in what way is the final statement a reminder of what and how we investigated a problem?

The notion of *genealogy* should provide an excellent metaphor for us to re-examine a number of issues in the learning and teaching of mathematics in addition to the context of elegance. We shall give the reader an opportunity to suggest dimensions to this notion before posing the question in more general terms in the last section of III. Let us now take a peek at one further possible dimension to elegance that the notion of genealogy suggests — namely its opposite — *potential*.

## Potential:

We discussed some spin-off that resulted from the deep proof of the sum problem towards the end of Section II. When I discussed it in class, many configurations and algebraic counterparts in addition to those depicted were suggested by my students on the basis of the staircase model. All of them however implicitly assumed that the staircase had to be "regular" shaped — like

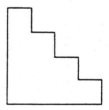

After a fashion, one member reluctantly suggested the following:

The suggestion was followed by the following analysis:

*Geometric Configuration*                                    *Algebraic Analysis*

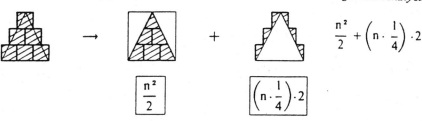

$$\frac{n^2}{2} + \left(n \cdot \frac{1}{4}\right) \cdot 2$$

I found a kind of elegance here that is essentially a "set breaker." That is, this *approach* (regardless of the success or failure of it in answering the question posed) opened up many avenues of exploration beyond the staircase varieties. Furthermore these directions not only encouraged us to analyze the original question—but *led us to pose new ones that no one had anticipated before.* For example, people began to investigate spiral staircases, wrap-around staircases and other varieties as well. That is, this *approach* was elegant in that it indicated greater *potential* for the geometric musing than any of us had been aware of. It opened up the question of what our thinking might become as well as where it came from. I am very much intrigued by the possibility that we could use this kind of analysis in order to encourage reflection both in affective and cognitive domains with regard to our own emerging thoughts and feelings. Furthermore we might use examples of this sort for the purpose of discussing where in a personal sense our priorities lie with regard to genealogy and potential.

How are we (and should we be) led to think about the present on the basis of connections with the past as well as of glimpses of possible futures? To what extent can and should we use mathematics of this sort as a springboard to reflect on such issues is a problem that I do not believe has been posed very seriously by anyone in the business of designing mathematics curriculum.

## IV. INTERMEDIATE SUMMARY, DISCLAIMER, A NOTE ON SANITY AND SIGNIFICANCE

With regard to elegance in Section III we have pointed out that the following sub-categories require a closer analysis:

1. elegance of proofs
2. elegance of assertions
3. elegance of interaction between assertion and proof
4. elegance of approach

In addition to rather obvious dimensions we might want to consider in analyzing these sub-categories of elegance, we have pointed to two that are perhaps less obvious: that of *potential* and *genealogy*. Furthermore these two dimensions are ones that ought to bear closer analysis than we have suggested even outside of the realm of elegance. In most general terms we might ask: what testimony is appropriate to offer to the fact that some significant thinking has taken place? How can we tell when we have tamed an idea long enough to know that it is capable of multiple births? We have neither attempted to analyze very seriously the notion of elegance nor have we gone far in clarifying the above two questions that are essentially questions of genealogy and potential. I hope this will be a start in both directions.

A final note on sanity with regard to the previous two sections would be in order. We have interspersed analysis with some description of classroom interaction. It is perhaps worth stressing that the students were colleagues or graduate students with a broad interest in education together with pre-service teachers, and many of the ideas in Sections III were generated during class discussions. I am *not* advocating the direct transference of the ideas generated herein in the precise form they have been suggested for a younger audience. I question seriously the extent to which many of these ideas ought to be verbalized in the explicit way we have been suggesting here though I believe the issue is open and represents fertile ground for serious research.

There are a number of kinds of ways I would think, however, in which an analysis of the content and issues of II and III could be helpful both to the classroom teacher and the curriculum designer. Let me suggest two non-obvious possibilities for each audience: (1) pedagogic propadeutic, and (2) humanistic links. (1) At one end of the spectrum our activity could be viewed as a pedagogic propadeutic. That is, though one might not wish to engage students directly in this kind of activity (search for surface, deep and formal connections; attempt to make sense out of notions such as genealogy and potential), an analysis along these lines multiplies the options from which one might then make selections in an attempt to teach a topic such as that of finding a succinct way of expressing the sum of the first n integers. The exercise of attempting to find a correspondence between algebraic and geometric formulations of a problem might be helpful in pointing out a geometric model (for example) that one had not previously considered.

Of course nothing in this kind of behavior on the part of the teacher would in itself prescribe classroom action. Just increasing possibilities does not inform one of the options he *ought to* consider in attempting to teach it. Value judgments with regard to what is worthwhile and psychological variables must also be considered.

Most teachers who tend to emphasize psychological variables focus on those that are primarily cognitive in nature. Some sensitivity to psychoanalytical and affective considerations is not incompatible however with some of the musings in the past two sections. For example, some of the models (see especially the "spin off" section of II and the "potential" section of III) involve the addition of missing pieces and some the elimination of what is too much.

It would be intriguing to find out how (and if) gender relates to these two kinds of explanations. Who is threatened by a removal explanation? Who tends to "add on"? At very least such considerations might sensitize us to an area we tend to ignore completely on mathematics education—namely gender differences.

(2) At the other end of the spectrum we could look at the kind of activity we described herein as a first approximation in a serious search for links between mathematics and realms of thought that are normally considered more personal and humanistic. In what ways could (should) we design curriculum around a self conscious attitude towards what we are thinking as well as towards the heuristics of thought? To what extent do the distinctions regarding similarity and difference in Section II get at important aspects of thought applicable in other domains? How might we carry the notions of genealogy and potential further in thinking about mathematics learning and how are they applicable in other realms as well?

We ought to be clear that the issue here is not one of transfer of training as normally conceived, nor of faculty psychology which was beaten to death years ago. I am certainly not claiming that doing mathematics will train the mind to think logically in some general sense. I am asking rather whether reflection about mathematical experiences might be used as a reasonable starting point (in conjunction with other areas) in order to encourage a kind of self consciousness that is frequently ignored especially by those scholars who intentionally select that discipline as a means of avoiding the kind of confrontation with self I am suggesting.

We turn in the next section to one more look at the manner in which mathematics might be used for the purposes we have in mind. I hope readers will bear in mind the kinds of disclaimers we have made in this section on sanity in reading what follows and that they will also be encouraged to pose related problems that carry the analysis beyond what I am suggesting.

## V. LEVELS OF EXPERIENCE

For many people an ability to clearly state goals beforehand is synonymous with the possibility of success in an educational enterprise. This is certainly the case with those who approach education from the point of view of behavioral analysis, but exists also in those whose approach is "softer."

Having learned to fashion education after technological models, we come very quickly and easily to ask ourselves and our students questions such as:

What am I trying to accomplish?
How can I get there?
What evidence will I have that I have arrived?[4]

Whether these are questions we keep well hidden in the back of our minds or are questions we ask explicitly at every stage of development of an educational enterprise, we find that many educators are caught in an essentially engineering conception of education.

Atkin (1970) has written an excellent criticism of the logical and psychological drawbacks that result from the extreme example of requiring an early statement of goals in behavioral terms. His main thesis is that such a prerequisite unduly limits what we will see and how we will see it, and that in any sufficiently rich educational enterprise, there will always be greater potential than we can envision at the beginning. Schwab (1969) goes further to point out the blindness that results from attempts to define curriculum problems at early stages from a limited number of theoretical perspectives regardless of their behavioral orientation. He argues well for a clear statement of objectives as a *last* step at best in curriculum design and one which testifies to the fact that deliberation has occurred.

How might we apply these criticisms to mathematics and in particular to the example we have been discussing in this paper?

## A Focus on "It":

Look once more at what we have done in Section II. Though we have suggested schemes that might have wide applicability (search for surface, formal and deep similarities), and though in doing so we have viewed our problem from many different perspectives (no mean feat for those who view the discipline as essentially from the point of view of singular approaches), we have very much focused on *the goal* — namely to find a more compact way of expressing $1 + 2 + 3 + \ldots + n$.

Now in one sense, what alternative do we have? This is the problem that was posed and so we view our task as one of answering it and pat ourselves on the back for approaching it in as imaginative a way as possible. Let us look at one alternative approach and then generate educational questions

---

[4]These questions appear on the surface to draw some similarity to those posed with regard to potential and genealogy in the previous section. The resemblance, however, is only a *surface one* since the questions are asked in a metaphorical, existential sense in the previous section, while here they are usually meant to be very real, hard and literal.

based upon other options and basic assumptions that this approach suggests.

## A Loss of Focus:

Look once more at the expression:

$$1 + 2 + 3 + 4 + 5 + 6 + 7 + 8 \ldots + 96 + 97 + 98 + \quad 99 + 100$$

We can make a number of obvious and not so obvious observations. For example:

1. The terms alternate between even and odd numbers.
2. We can form a new series by adding pairs of terms:

$$1 + 2 = 3, 3 + 4 = 7, 5 + 6 = 11, \ldots$$

We thus have:

$$3 + 7 + 11 + 15 + \ldots$$

The difference between terms is now 4.

3. We can form a new series by adding pairs of consecutive terms by allowing repetition of an element as we did not allow in (2):

$$1 + 2 = 3, 2 + 3 = 5, 3 + 4 = 7, 4 + 5 = 9 \ldots$$

We get a new series:

$$3 + 5 + 7 + 9 + \ldots,$$

which is just the sum of all odd numbers greater than 1.

4. If we subtract consecutive pairs in the spirit we added them in (2), we get:

$$1 + 1 + 1 + 1 + 1 + \ldots$$

We have barely scratched the surface but have done enough to suggest that the observations one might make in this context are almost without limit. Furthermore, we have observed aspects of the original phenomenon that appear to have some regularity. We might just as profitably have observed irregularity.

## Out of Focus Brought In:

One rather obvious extension of the above activity would be to use it to gain some insight into the original question posed.

Let us show *one* successful attempt.

Look at (1): the terms alternate between odds and evens. We might further observe that they do so in such a manner as to expose consecutive odds and evens. Suppose we separate the odd and even terms of the series. What do we have?

$$1 + 3 + 5 + 7 + \ldots + 97 + 99 \text{ and } 2 + 4 + 6 + \ldots + 98 + 100.$$

How might we now recall a goal that was established at the beginning? One possibility now would be to re-define the goal with the hope that re-definition would lead to a new synthesis. As a start we might begin to re-define the goal. Suppose we deal separately with the odd and even sum sections. For the odd, the following picture is suggestive if we use squares as in the Wertheimer case:

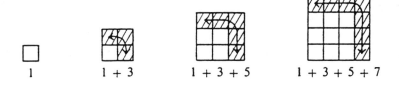

| 1 | 1 + 3 | 1 + 3 + 5 | 1 + 3 + 5 + 7 |

Since we can find a geometric scheme which involves always going from one square to another it is not hard to see why it is that adding together the first n odd terms gives a sum of $n^2$. But what scheme of a geometric nature is suggested by the even sum? The following picture again is helpful (though there are certainly other options).

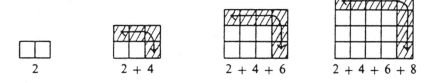

| 2 | 2 + 4 | 2 + 4 + 6 | 2 + 4 + 6 + 8 |

Look at the dimensions of each of the rectangles. The first is $1 \times 2$; the second $2 \times 3$, the third $3 \times 4$ etc. In general we might not find it hard to persuade ourselves that the $n$th sum would yield a rectangle with dimensions $n \times (n + 1)$. How do we now try to re-combine our two new observations about odd and even terms? The following scheme makes sense if we deal with the series whose last term is even:

$$1 + 2 + 3 + 4 + \ldots + 96 + 97 + 98 + 99 + 100:$$

$$(n - 1) \text{ term} \qquad n^{th} \text{ term}$$

For the odd sum we get:

$$\left(\frac{n}{2}\right)^2$$

That is, if n = 100, to get the sum of the odd terms through 99, we find out how many terms there are and take the square of the number of terms. So, since the number of odd terms is

$$\left(\frac{99 + 1}{2}\right) \text{ or in general } \left(\frac{(n - 1) + 1}{2}\right)$$

we find the sum to be

$$\left(\frac{100}{2}\right)^2 \text{ or } \left(\frac{n}{2}\right)^2 \text{ in general}$$

For the sum of the even terms, using the rectangle configuration we find the sum to be

$$\left(\frac{100}{2}\right) \times \left(\frac{100}{2} + 1\right)$$

That is it is of the form

$$\left(\frac{n}{2}\right)\left(\frac{n}{2} + 1\right) .$$

If we were inclined to "simplify," we would find that adding the odd and even sums "reduces" as follows:

$$\left(\frac{n}{2}\right)^2 + \left(\frac{n}{2}\right)\left(\frac{n}{2} + 1\right) = \left(\frac{n}{2}\right)\left(\frac{n}{2} + \frac{n}{2} + 1\right) = \left(\frac{n}{2}\right)(n + 1).$$

The right hand side is a familiar form. Consider the case of an odd number of terms,

$$1 + 2 + 3 + 4 + \ldots + 97 + 98 + 99$$
$$(n - 1) \, n$$

This scheme suggests

$$\left(\frac{n + 1}{2}\right)^2 \text{ for the sum of the odds and}$$

$$\left(\frac{n - 1}{2}\right) \bullet \left(\frac{n - 1}{2} + 1\right) \text{ for the sum of the evens.}$$

Again "reducing," we get:

$$\left(\frac{n+1}{2}\right)^2 + \left(\frac{n-1}{2}\right)\left(\frac{n+1}{2}\right) = \left(\frac{n+1}{2}\right)\left[\left(\frac{n+1}{2}\right) + \left(\frac{n-1}{2}\right)\right] = \left(\frac{n+1}{2}\right) \bullet n.$$

The right hand side is another familiar form, different of course from the previous one.

### Reflections On The Previous Two Subsections:

Instead of attempting to answer the question posed directly, we began by making observations on the thing that was put before us to investigate. After some musing, we chose one of the observations—(1)—and used that as a new starting point to once more get at the old goal. We did not in any significant sense revise the goal; we merely postponed it and returned to it at a later date.

Let us consider other ways in which we might either postpone or modify a goal that was put to us at the beginning:

(A) One obvious activity would be to follow the musings of (1)-(4) with an internal *logical analysis.*

How do these four statements relate to each other or how do they relate to the original problem? Are they logically equivalent? Does one imply the other? Is one more primitive in some sense than the other? If we demonstrate logical equivalence (or perhaps that one of the other observations is even stronger than what we were asked to search for originally), we could with the same goal in mind work on one of these assertions rather than the original question. In addition to a logical analysis connecting up all other observations, we might ask which of them (regardless of logical connections to the original question) are true? How might we find out? Which are false? Which fixable?

(B) A more radical activity might be to modify the original problem considerably. That is, instead of asking: what is the sum of the first 100 integers, or asking: what observations we might ask based on this series in particular, we might do any of the following:

1. Suppose we change sum to product.
   What is: $1 \times 2 \times 3 \times \ldots \times 97 \times 98 \times 99 \times 100$?
   What observations can we make on the series?
   How does it differ from the original?
2. Suppose we focus only on terms that are multiples of 3?
   What do we observe?
3. Can we find a three dimensional analogue for the staircase model?

That is, we could use the original question not only for purposes of making observations on it that could later be reconnected to the original goal, but we could use the original problem as an inspiration for setting new goals. Here we begin to lose sight of the original problem, though it is possible that such activity might (among other things) make clearer the object of our original focus.[5] Both of these above modes of operating, are an attempt to either clarify or modify a world out there. Though mind expanding in that the activity places value on perspectives other than those that were imposed from the outside, there is so far very little influence on the "self" as an object of study. These kinds of activities are justified however on grounds that they are means of making something one's own. Let us not minimize such an enterprise. Activity such as

*making observations of a phenomenon presented*
*drawing implications from these (assumed valid) observations*
*using the phenomenon to imagine alternatives to it*
*negating some of the hypotheses*
*posing new problems*

can be thought of as a means towards incorporating an *abstraction* that is "out there" in such a way that we begin to gain power over it and to feel that we possess it in some important sense. This kind of "tasting" activity is one that we tend to by-pass if our focus is primarily upon solving "it." Unfortunately, however, if we persist in by-passing this activity very long, we desensitize ourselves to the point that we no longer "taste" the uniqueness among phenomena, and though they may be able to gain answers to questions, they become very much insensitive to what it means for something to be a problem and have even less of an understanding of what it means to have solved something.

Make no mistake about it. Our strategies suggested under (B) are revolutionary ones for those whose major concern in curriculum is to hand over a hermetically sealed body of knowledge. As suggested in much of this article up to this point, in order to acquire the personal incorporation of the kind we have been illustrating with just *one* example requires both considerable time and freedom to dawdle (and to be unsuccessful). The great irony is that I am certain most scholars who advocate the covering of material in a way that would look respectable to their colleagues in the field have themselves made frequent type (B) incorporations, though they may be either unaware or have forgotten that they have done so. Some of the

---

[5]It has become clear to us that semi-systematic modification on things being investigated followed by posing of questions on the modified form (occasionally superimposed on the original) can be one way of getting a clearer idea of what is being investigated.

components we have summarized above are part and parcel of at least the less than conscious mode of anyone who begins to truly understand something from the inside rather than as a message from without. Through allowing of *deviations* from the original "problem" and establishment of new goals (either consciously or unconsciously—if that is possible) one may begin ironically enough to become clearer on what the original phenomenon is about. Even if this does not result however, activity of the kind we have described should provide at least a partial condition for making *something* (difficult as it may be to specify what that something is) one's own rather than appreciating it as an outsider. In many cases even a bow to learning by discovery overlooks the kind of incorporation we have been suggesting. See Brown (1971)

Despite the radical tone of what we have been suggesting in (B), we should be aware of the fact that the focus is still primarily upon "digesting" in an adequate and imaginative manner an outside force.

In what ways might we go beyond (B) in an attempt to gain some deeper understanding of self? What has characterized all the approaches in both (A) and (B) so far has been an attitude of *exegesis*. That is in both approaches, though the activity has certainly not been an attempt at by-passing one's mind as in attempts of conditioning, there has been an implicit acceptance of the phenomenon under investigation. In (A) though we temporarily postponed an original goal imposed, we eventually played the game of synthesizing results so as to return to what was called for. In (B), though we strayed further, we accepted what was "out there" and attempted to modify it only in so far as we might better understand it by imposing variations of many sorts. How might we be less exegetical in using mathematics—and more generally the disciplines—in effecting self-awareness? Let me suggest several possible directions. What follows is merely a sketch and requires much deeper analysis than I am prepared to provide at this point. Pleasingly enough I will, at points, have to stray beyond the problem of finding an expression for the sum of the first n consecutive integers at this point, since my thinking is not well enough defined to make sense in the more or less isomorphic way I have attempted so far.

## VI. BEYOND EXEGESIS

In Section III we pointed towards the possibility of using mathematics for the purpose of reflecting on potentiality and genealogy as these dimensions affect us in an existential way beyond that discipline per se. We might use this idea as a first approximation in going beyond an exegetical stance towards the disciplines.

Even in the case of potential and genealogy we sanctify the subject by requesting that it be used as a trigger for incorporation of humanistic themes. Perhaps a more radical question we might pose could be:

Should I accept what this example implies as a means of knowing? experiencing?

More generally, what is worthwhile (for me) about the disciplines as a means of coming to know or experience? What do I find out about myself in using examples from the disciplines? What are alternative ways of experiencing?

Though I assume a rational approach in coming to answer these kinds of questions, I do not see pre-ordained answers for the individual, nor do I imply that the disciplines provide the only experience possible for beginning to answer these kinds of questions. In genuinely asking the *acceptance* question, I am allowing for rejection of a discipline or an aspect of a discipline as a method of coming to an awareness of self. In looking at the question from the perspective of the emerging self, I am furthermore *not* assuming that it makes sense to ask these kinds of questions only from a detached disinterested point of view.

Peters (1967), in asking the questions such as "What activities are most worthwhile?" or "Why do this rather than that?" assumes that a kind of detachment and disinterest are prerequisites for a serious consideration of these questions. Though one might certainly wish to be governed by more than a pleasure principle in answering these kinds of questions, I do not see the incompatibility between asking them in a rational sense and at the same time from a very personal, idiosyncratic point of view in which one attempts to invest many aspects of one's own uniqueness in answering them. As a matter of fact as soon as we attempt to answer such questions beyond gross partitioning (that is in a more minute level than: "Are the disciplines more worthwhile than activities such as bowling?"), it is probably logically impossible for one to assume the kind of detachment Peters advocates.

In order that rejection (or acceptance for that matter) of disciplines or parts of disciplines and subsequent self actualization not be a totally superficial activity, there would certainly have to be *some* attempt made to view the disciplines from within—or to be initiated into them as Peters suggests (see pp. 165–169). But the object for the purpose of self actualization would be to find some balance between the tensions of involvement (or commitment) and withdrawal (assuming an attitude of tentativeness).

The job of teaching by use of the disciplines is much more complicated than Peters suggests if our aim is self actualization. One has to balance commitment to truth as expressed within a body of knowledge or emerging knowledge, with an attitude of concern for how that truth sheds light in an idiosyncratic way on the emergence of a self. As we move to self, the truth "out there" may be nothing more than a mirror against which one reflects his own image and emerges with new personal insights. There may, in fact, result logical incompatibility between the truth out there and what emerges within. We shall illustrate shortly what this means and how it might result.

What are some dimensions to the discipline of mathematics that might help one to emerge with greater self understanding? It seems to me that at very least one must find ways of providing both more significant *choice* and more significant *doubt* within the disciplines in order that self actualization not be a shallow emergence. Let us look at various dimensions of choice first.

## Dimensions of Choice:

We have already indicated some aspect of choice in the example of the sum of the first n integers. One might stand back from his experience and ask himself in purely cognitive terms how he is affected by various aspects of the visual and the abstract. Specifically how does an algebraic vs. a geometric approach to this problem prompt one to reflect on his own modes of viewing the world? What kinds of images inspire him or aid him to understand what aspects of his environment?

Of the many different approaches to the problem some are more plodding and some more elegant. In addition to some of the questions we have already suggested with regard to elegance, if one is exposed to or encouraged to generate an abundant number of elegant and inelegant approaches, then new questions and categories regarding individual styles of operating may emerge. What are some of the important variables that affect the ways in which we operate with regard to an elegant vs. a plodding approach to problems that exist around us? To what extent are we affected by self-consciousness? clarity of problem? sources of initiation of a problem? One could go on asking questions of this sort. Furthermore, I believe that the relative cleanliness of the issue within the domain of mathematics may provide both *the initial distance* and the mirror to inspire one to view the issue in more personal and general terms. Another dimension of choice that we have already alluded to deals with the distinction between posing and solving problems. We have become accustomed to thinking of mathematics as an enterprise involving our attempts to solve, but rarely to pose problems — though in a most basic sense, it is probably impossible to solve a problem without posing some related ones along the way. Once we have gained some experience and been exposed to strategies of posing problems, how can we use it as a springboard for reflecting on ourselves as posers or solvers? How do we tend to pose problems? Under what circumstances do we do so? In what ways does our vacillation between solving and posing in mathematics suggest alternatives or reflect itself in the way we operate in other spheres as well? How does our ability or inability to pose problems (or our inclination or disinclination to do so) tend to define us with regard to notions such as authority (who's in charge?), self respect and the like? No doubt the reader can expand both the discussion of choice and the kinds of questions that might lead to self actualization.

## Doubt:

With increased choice comes the possibility of greater doubt through conflict. What happens when we are exposed to radically different conclusions based upon different approaches to the same problem? There are many ways in which this may come about, and certainly the analysis of apparent paradoxes is an extreme example of it. We frequently tease students with proofs that every triangle is isosceles, that a right angle is greater than 90°, that 2 = 1 and so forth.

Though paradox is an extreme case, there are many instances where conclusions X and not X may be in tension with each other though not logically inconsistent. I have written elsewhere for example of the nature of the intersection point formed by two lines in the plane, each of which has two rational points (where a rational point in the plane by definition has two rational coordinates). Algebraically, it is highly expected that the intersection itself will be a rational point. Viewed from the perspective of chance however, the probability that this would come about (even knowing the algebraic analysis) is 0.[6]

One source that may lend itself to greater doubt through conflict may come from the disparity between intuitive and rigorous approaches to the same problem. Consider the following for example:

Place a belt around the earth. Now add 30 feet to the length of the belt and again place it around the earth — so that the belts and the earth share the same center. Consider the rim around the earth left by the larger belt. What would fit between it and the earth? An ant? A pin? A marble? An elephant?[7] Make an intuitive guess of what the answer would be — taking into consideration that 30 feet is almost nothing compared to the circumference of the earth. Compare your intuitive judgment with a more rigorous analysis. What do you conclude?

How does one adjudicate radically different answers to the same prob-

---

[6]A zero probability does not imply impossibility in the case of probability for the infinite. So though things are unexpected, they are logically possible. See Brown (1971a).

[7]For a discussion of the conflict of intuition and rigor within the context of moral education, see Brown & Lukinsky (1970).

lems? Where does he look for resolution? Under what circumstances will he tolerate conflicting conclusions? One can begin to use mathematical examples of this sort to once more broaden the range of these kinds of questions to include thought and action that go beyond mathematics and in fact, go beyond the disciplines. An interesting aside here is that even within the community of mathematics scholars, it seems not to be the case that the most accepted of competing approaches to a problem are those that are the most rigorous. Surprisingly there have been instances in which the deciding factor has been the degree to which rigorous approaches accorded with former intuitions rather than the other way around. See Rogers (1964). If this is legitimate within mathematics, how much more so might it be legitimate within personal realms as well?

It seems to me that in much of our educating, we attempt to by-pass the kind of doubt and conflict I have been suggesting in this subsection. This is especially so if our object is to make things intuitively plausible. Especially from this perspective, we ought to take a second look at the uses and purposes of concrete materials in mathematics education. It is not necessarily the case that their only function is to make things more "real" and obvious, but to the extent that that is the purpose for which they are being used, we should appreciate what it is we may be sacrificing.

I am suggesting that the inculcation of doubt may be the first step in any serious investigation of mathematics for its own sake as well as for the sake of triggering thought in other realms as well. If we seriously believe this to be the case, then where do we begin? What are the faces of doubt in mathematical thought?

## VII. SUMMARY

Hopefully the reader has by this point experienced intense doubt with regard to some arguments of the previous sections. Such doubt most certainly must be based at least upon observations of ambiguity, and inconsistency. If what I have claimed in the previous subsection however, has any validity, then we are both on the road to worthwhile further enlightenment.

This program is clearly a beginning and I look forward as much to readers' clarification and extensions of some of these ideas as I do to the less exegetical activity of condemning them.

I particularly welcome criticism that focuses on the role of mathematics per se in what I have said. For which of my claims, could one just as easily substitute any other discipline? I suspect this may be the case "almost everywhere" in the paper, but I am not sure. While such criticism would be helpful, I do not believe the claim of nonuniqueness would damage most of my observations. On the contrary, it might inspire us with even greater optimism if what most people conceive to be among the coldest of all disciplines, could shout "me too" with regard to claims of linkage to humanistic themes.

We alluded in Section I to negative psychological or epistemological attitudes frequently conveyed to students in the study of mathematics. Perhaps at this point it would be worth spelling them out in isolation from the alternatives with the hope that the reader might wish to further criticize these assertions as well as come up with his own set of alternatives where appropriate.

Some of the attitudes are well known and several of the recent mathematics programs have attempted to counteract their force.

For example:

1. It is necessary to have an outside authority to judge correctness of any alleged answer to a mathematical problem.
2. There is essentially one way of solving any mathematical problem.
3. One's job is to learn what the rules are and how to follow them. Deviation from them constitutes a moral sin.

We need some hard headed research to determine the extent to which such attitudes acquired in one domain (such as mathematics) affect one's *conception of knowledge* (and not merely what he believes or knows) in others.

There are many other harmful but more subtle attitudes that I believe are conveyed in the context of much of mathematics learning and I do not believe we have begun to see these quite so clearly as the first set:

4. All problems should come clearly posed or they are either incapable or unworthy of being analyzed.
5. One's job is to *solve* problems if possible. Posing belongs to the realm of the experts.
6. In solving problems, one's focus ought to be constantly on the questions being asked.
7. The problems stand "out there," to be solved, and are not capable of shedding much light on the problem of understanding oneself.

To what extent do the options presented here generate alternative conceptions of knowledge, and in what sense are these humanistic conceptions reasonable alternatives?

We should stress that we have used "humanistic" in a very broad sense, but that we have certainly not exhausted the possibilities even within that broad sense. At one end of the spectrum (Section II) we have been concerned with heuristics for thinking that may have general applicability and that go beyond common distinctions of thought such as induction vs. deduction. Towards the middle of the spectrum we have asked what kinds of things one might do in order to internalize what is "out there" (Section V). At the other end of the spectrum, we have explored the possibilities of using the "out there" for the purposes of self actualization. Our discussion of genealogy and potential in Section III was a start, but Section VI, BEYOND EXEGESIS suggests how mathematics might be used as a trigger for self actualization in a deeper sense.

## REFERENCES

Atkin, J. M. (1970). Behavioral objectives in curriculum design: A cautionary note. In J. Martin (Ed.), *Readings in the Philosophy of Education* (pp. 32–38). Boston: Allyn & Bacon.

Brown, S. I. (1965). Of *prime* concern: What domain? *Mathematics Teacher, 58,* 402–407.

Brown, S. I. (1968). *Prime* Pedagogical Schemes. *American Mathematical Monthly, 75,* 660–668.

Brown, S. I. (1971). Learning by discovery in mathematics: Rationale, implementation and misconceptions. *Educational Theory, 21,* 232–260.

Brown, S. I. (1971a). Rationality, irrationality and surprise. *Mathematics Teaching, 55,* 13–19.

Brown, S. I., & Keren, G. (1972). On problems in gestalt psychology and traditional logic: A new role for analysis in the doing of mathematics. *Journal of Association of Teachers of Mathematics of New England, 5*(1), 5–13.

Brown, S. I., & Lukinsky, J. S. (1970). Morality and the teaching of mathematics. *Ethical Culture Society, 1*(2), 2–12.

Goodman, N. (1972). *Problems and projects.* Indianapolis: Bobbs-Merrill Co.

Peters, R. S. (1967). *Ethics and education.* Glenview, IL: Scott Foresman & Co.

Rogers, R. (1964). Mathematical and philosophical analysis. *Philosophy of Science, 31,* 255–64.

Schwab, J. J. (1969). The practical: A language of curriculum. *The School Review, 78*(1), 1–23.

Wertheimer, M. (1945). *Productive thinking.* London: Harper Brothers.

# The Intertwine of Problem Posing and Problem Solving: Editors' Comments

The two essays in this section explore an array of connections between problem posing and problem solving. In the first essay, **Problem Posing in Geometry,** Hoehn takes a rather innocent looking theorem from geometry:

> If two tangents are drawn to a circle from an external point, then the two tangent segments are congruent.

He then creates no less than thirteen variations of the problem based upon an array of problem posing heuristics. What is interesting about these heuristics however is that many of them are the same as heuristics we associate with problem *solving*. In addition to the **looking back** heuristic explored in the previous section, he looks at patterns, special cases, generalizations, converses, related problems and so forth.

Some of these heuristics are included in what we describe as "accepting the given" in Chapter 3 of **The Art of Problem Posing** (for example, searching for patterns). Some are part of the What-if-Not scheme in which the given is challenged. These matters are discussed in Chapter 4 of **The Art of Problem Posing.** For example, to look at the converse of a proposition is to do a What-If-Not on the *structure* of the statement rather than its meaning ("The statement is a proposition of the form P implies Q. What if it were not a proposition of that form? What could it be? . . .). Despite the fact that some of Hoehn's heuristics can be reduced to a previously discussed What-If-Not format, his observation is an original insight that has potential power to enable us to see connections we may never have realized before. It is worth exploring problem solving heuristics in general to see the extent of their applicability to the problem posing scene as well. And of course . . . vice versa.

One should appreciate that many of the problems created by Hoehn do in fact require a fair amount of knowledge about the subject. It would be unusual for someone who has barely learned the theorem Hoehn explores to pose questions of his level of sophistication. Nevertheless, the categories he invokes are ones that would enable even the neophyte to create an interesting array of new problems.

In the second essay in this section, **Students' Microcomputer-Aided Exploration in Geometry,** Chazan explores the potential of the Geometric Supposer as a device to enable inquiry among middle school and high school students. Because measurement of lengths, angles and areas is built into the program, there is considerable opportunity for students to operate empirically as a precursor to a more formal approach.

Rather than exploring common features between problem posing and problem solving as in the previous article, this piece views several components of the web of inquiry and explores the ways in which they depend upon each other. Among the elements of the web are: verifying, conjecturing, generalizing and posing new problems, communicating, proving and making connections.

With regard to conjecturing and to problem posing, Chazan suggests ways in which teachers might encourage the process. He points out that in addition to providing students with strategies for generating their own problems (like focusing on a subset of a system or like searching for relationships among members of a class), they have to be made aware of the fact that such activity is both possible and valued. He has a number of suggestions for the kind of activities teachers can perform to encourage that activity.

There is some interesting overlap (using different language) between suggestions made by Chazan and by Brown and Walter for encouraging the posing of new problems even prior to a What-If-Not point of view. As you read his article, you might wish to see the ways in which it connects up with and cleverly applies some of what we call *Phase One Problem Generation* in Chapter 3 of **The Art of Problem Posing.**

# 27 Problem Posing in Geometry

Larry Hoehn

Problem solving received dramatic emphasis throughout the 1980s. However, its necessary counterpart, problem posing, has received scant attention. Some notable exceptions are the works of Brown and Walter (1983), Klamkin (1986), and, of course, Pólya (1973). In this article a typical geometry theorem is used and problems are posed involving its application. The methods presented here work well to create geometry test questions, geometry contest problems, and calendar problems for the *Mathematics Teacher*. Although these methods are intended primarily for secondary school teachers, creative geometry students could use them for making their own mathematical discoveries.

First suppose that we have completed study of a unit on circles. We then wish to test our students' understanding of the following theorem:

THEOREM. *If two tangents are drawn to a circle from a point outside the circle, then the two tangent segments are congruent.*

The usual proof of this theorem (see Fig. 27.1) is to prove $\triangle PAO \cong \triangle PBO$. An alternative proof, however, is to show that $m < PAB = (1/2)m(AB) = m < PBA$. Then $\triangle APB$ is isosceles and thus $PA = PB$.

We can certainly present all kinds of numerical exercises to test students' understanding of this theorem (and using some are fine), but we prefer problems that are more geometrical than arithmetical. In other words, we want students to use the theorem to prove a related result.

One way of obtaining "related results" is slightly to increase the complexity of the figure. For example, suppose we introduce another tangent to the circle such as shown in Fig. 27.2. Then we have a problem that can be

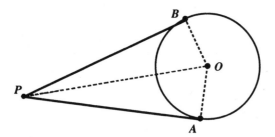

FIG. 27.1.Prove *PA* = *PB*.

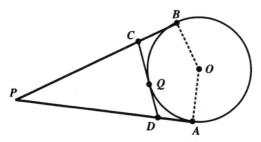

FIG. 27.2.   Prove *PA* + *PB* =
*PD* + *DC* + *CP*.

stated at two different levels of difficulty, each of which requires the same applications of the theorem for solution.

*Problem 1.* Let $\overline{PA}$, $\overline{PB}$, $\overline{CD}$ be tangents to a circle as shown in Fig. 27.2. Show that *PA* + *PB* = *PD* + *DC* + *CP*.

*Problem 2.* Let $\overline{PA}$ and $\overline{PB}$ be tangents to a circle and let *Q* be an arbitrary point on $\overarc{AB}$. If $\overline{CD}$ is tangent to the circle at *Q*, then the perimeter of $\triangle PCD$ is constant.

A second way to increase the complexity of figure 27.1 is to add another tangent to the circle but in a different position, such as shown in Fig. 27.3. Two proposals, but again at quite different levels of difficulty, immediately come to mind:

*Problem 3.* If *A*, *B*, and *C* are the points of tangency of sides $\overline{PQ}$, $\overline{QR}$, and $\overline{RP}$ of $\triangle PQR$ to a circle, then *PA* + *QC* + *RB* = *AQ* + *CR* + *BP*.

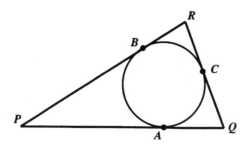

FIG. 27.3.   Prove *PA* + *QC* +
*RB* = *AQ* + *CR* + *BP*.

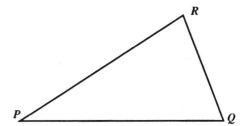

FIG. 27.4.  Construct *X*, *Y*, and
*Z* such that *PX* = *PZ*, *QX* = *QY*,
and *RY* = *RZ*.

*Problem 4.* Construct points *X, Y,* and *Z* on the sides of an arbitrary △*PQR* (see Fig. 27.4) such that *PX* = *PZ*, *QX* = *QY*, and *RY* = *RZ*. The solution is to inscribe a circle in △*PQR*. Figure 4 suggests another method of posing problems—simplify a figure (i.e., Fig. 27.3). This simplification forces the problem solver to come up with the missing points.

A third method to create problems is to combine two or more results. Here we wish to combine Fig. 27.2 and Fig. 27.3 to obtain the following:

*Problem 5.* Let △*PQR* and hexagon *DEFGHI* be circumscribed about the same circle as shown in Fig. 27.5; then △*PQR* has the same perimeter as the combined perimeters of △*PDI*, △*QFE*, and △*RHG*. The proof is three applications of problem 1.

A fourth method of creating mathematical problems is to generalize a previous result. For example, to generalize problem 3, we begin by changing from a triangle circumscribed about a circle to a quadrilateral circumscribed about a circle. Thus, we see from Fig. 27.6 that *PW* + *QX* + *RY* + *SZ* = *WQ* + *XR* + *YS* + *ZP*. We then change the wording to "*n*-gon circumscribed about a circle" to get comparable results. A quadrilateral, however, is especially interesting, since the facts that *PW* + *QX* + *RY* + *SZ* = *PW* + *WQ* + *RY* + *YS* = *PQ* + *RS* and *WQ* + *XR* + *YS* + *ZP* = *QX* + *XR* + *SZ* + *ZP* = *QR* + *SP* imply that *PQ* + *RS* = *QR* + *SP*. This result interestingly poses another problem:

*Problem 6.* If a quadrilateral is circumscribed about a circle, then the sums of the lengths of the pairs of opposite sides are equal.

Some pitfalls are encountered in generalizing problems. Suppose we try to generalize problem 4 for quadrilaterals.

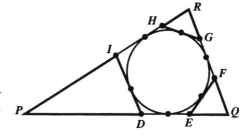

FIG. 27.5. The perimeter of
△*PQR* is the sum of the perimeters
of △*PDI*, △*QFE*, and △*RHG*.

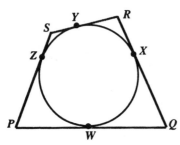

FIG. 27.6.   Prove $PQ + RS = QR + SP$.

*Problem 7.* Construct points $W, X, Y,$ and $Z$ on the sides of quadrilateral $PQRS$ such that $PZ = PW$, $QW = QX$, $RX = RY$, and $SY = SZ$.

Fig. 27.7 shows that problem 7 is not always possible to solve.

A fifth way to create mathematical problems is to consider the converse of a previous result. For example, the converse of problem 6 is as follows:

*Problem 8.* If the sums of the lengths of the pairs of opposite sides of a convex quadrilateral are equal, then a circle can be inscribed in the quadrilateral.

*Proof.* First we construct a circle inside the given quadrilateral such that it is tangent to three of the sides of the quadrilateral. Next we construct a second circled tangent to the fourth side and to two of the former sides. Two examples are shown in Fig. 27.8, where we denote the nonoverlapping segments by $a, b, c, d, e,$ and $f$ and the sides by $w, x, y,$ and $z$. Since $w + y = x + z$ in either figure, we have $(a + d) + (b + c) = (a + e + b) + (c + f + d)$. By simplification, $0 = e + f$. Therefore, $e = f = 0$. Thus, the "two circles" are identical, which completes the proof.

In the solution of problem 8 we considered two circles in a quadrilateral. What if the two circles are separated by a diagonal—perhaps both circles tangent to a diagonal? In problem posing we don't want to be too general at first, so we consider a rectangle such as shown in Fig. 27.9 with $x > y$. Since $m + c = b$, then $m = b - c = (a + b) - (a + c) = x - y$. That is, we have the following:

*Problem 9.* If circles are inscribed in the two triangles formed by a diagonal and two adjacent sides of a rectangle, then the length of the "midsegment" of the diagonal is equal to the difference of a pair of adjacent sides of the rectangle.

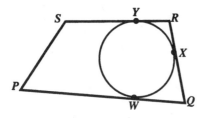

FIG. 27.7.   Construct $W, X, Y,$ and $Z$ so that $PW + QX + RY + SZ = WQ + XR + YS + ZP$.

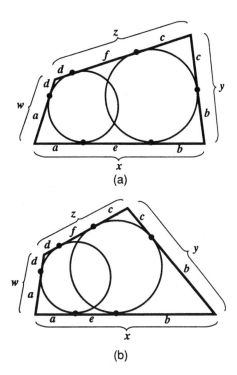

FIG. 27.8.   Prove $e = f = 0$.                                                (b)

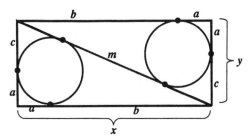

FIG. 27.9.   Prove $m = x - y$.

Since no properties of right angles were used in problem 9, we can replace "rectangle" by "parallelogram" to get what appears to be a new problem. Can we then replace "parallelogram" by "trapezoid" or some other quadrilateral? This question is asked in the next proposal.

*Problem 10.* Let $w, x, y,$ and $z$ be the lengths of the sides of an arbitrary convex quadrilateral as shown in Fig. 27.10. Find a formula for $m$ in terms of the sides $w, x, y,$ and $z$.

Using our main theorem about tangents, we have $e = f + m$ and $b = c + m$ (note that the figure assumes $e \geq f$ and $b \geq c$.) These equations can be rewritten as $m = e - f$ and $m = b - c$. Therefore $2m = (e - f) +$

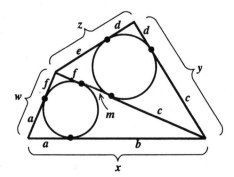

FIG. 27.10.   Prove $m =$
$$\frac{x + z}{2} - \frac{w + y}{2}.$$

$(b - c) = (e + b) - (f + c) = (e + d + b + a) - (f + a + c + d) = (z + x) - (w + y)$. Thus $m = (x + z)/2 - (w + y)/2$

Note that if $m = 0$ in Fig. 27.10, then $w + y = x + z$. Thus we have discovered another proposal "accidentally." Since accidents are by definition unplanned, we can't use this approach as a dependable method of problem posing; but we want to be alert for the opportunities (e.g., antifreeze was discovered accidentally). Our proposal is as follows:

*Problem 11.* If circles inscribed in the two triangles formed by a diagonal of a quadrilateral are tangent to each other, then a circle can be inscribed in the quadrilateral, and conversely.

By simplifying Fig. 27.10 we have another proposal:

*Problem 12.* If $\overline{AC}$ is the longest side of $\triangle ABC$, construct a point $D$ outside $\triangle ABC$ such that $AB + CD = AD + BC$.

First, we inscribe a circle in $\triangle ABC$ in figure 27.11. Let $P$ be the point of tangency with side $\overline{AC}$. At $P$ construct a circle outside $\triangle ABC$ but tangent to $\overline{AC}$. From $A$ and $C$ construct tangents to this circle and extend them to meet at some point $D$. If the tangents are parallel, then a smaller circle will ensure that they meet at $D$ as in Fig. 27.11, whereas a larger circle would

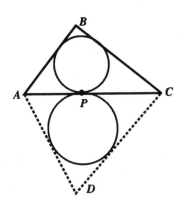

FIG. 27.11.   Construct $D$.

result in a nonconvex quadrilateral (with one exception). By problem 11, $D$ is the required point.

Another method of problem posing is to exploit the "symmetry" of a prior proposal. For example, in problem 10 and Fig. 27.10 we could have used the other diagonal of the quadrilateral (see Fig. 27.12) and similarly obtained $n = (x + z)/2 - (w + y)/2$. Thus $m = n$. Therefore, we arrive at the following:

*Problem 13.* Let 1, 2, 3, and 4 be the four overlapping circles inscribed in the triangles formed by a diagonal and two consecutive sides of a convex quadrilateral (as in Fig. 27.12). Let $m$ and $n$ be the lengths of the "midsegments" of the diagonals between the points of tangency of the circles to the same diagonal; then $m = n$.

As the reader can tell, the problems are getting more complicated. Therefore, the problem poser might want to work in other directions. For example, which of the foregoing problems have counterparts for —

1. nonconvex quadrilaterals;
2. convex or nonconvex pentagons, hexagons, and so on;
3. spheres and polyhedra;
4. the theorem "If from a point outside a circle two secants are drawn to the circle, then the product of one secant and its external segment is equal to the product of the other secant and its external segment"?

In posing these problems, the careful reader will note that we have used some of the same techniques that are especially useful for problem solving, for example, special cases, generalization, related problems, converses, symmetry, useful notation, accident, previous results, useful figures, looking back, patterns, and others. All that one needs to be a problem poser is first to be a problem solver. The two processes are inseparable.

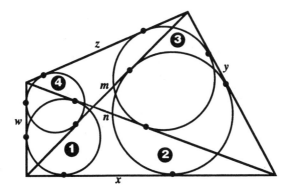

FIG. 27.12.   Prove $m = n$.

# REFERENCES

Brown, S. I., & Walter, M. I. (1983). *The art of problem posing.* Philadelphia: The Franklin Institute Press.

Klamkin, M. S. (1986). The Olympiad corner: 80. *Crux Mathematicorum. 12,* 263–281.

Polya, G. (1973). *How to solve it.* Princeton: Princeton University Press.

# 28 Students' Microcomputer-Aided Exploration in Geometry

Daniel Chazan

Four important themes presented in the K-12 *Curriculum and Evaluation Standards for School Mathematics (Standards)* (NCTM 1989) are mathematics as problem solving, mathematics as communication, mathematics as reasoning, and mathematical connections. The high school component also stresses mathematical structure. Furthermore, the *Standards* calls for new roles for teachers and students and suggests that microcomputer technology can help support teachers and students in taking on these new roles.

One genre of software that can help teachers accomplish these goals is tool software, which Schwartz (1989) calls "intellectual mirrors." Software of this kind does not ask questions and evaluate answers; it furnishes tools for users to test conjectures in a particular domain. On the basis of the feedback received from the software, a user can explore relationships among the objects in the domain and can examine their understanding of the domain.

This article describes the ways in which some teachers have used a particular piece of software of this kind, The Geometric Supposers (Schwartz, Yerushalmy, and Education Development Center 1985). (The Geometric Supposers consist of four pieces of software: the preSupposer [appropriate for middle school], Triangles, Quadrilaterals, and Circles). One geometric construction is used here to describe some of the ways these teachers have tried to help their students become better inquirers and to illustrate how the approach they have used addresses the central themes of the *Standards*.

## ONE WAY TO USE THE SUPPOSERS

The Geometric Supposers allow users to start with an "initial" shape (e.g., an acute scalene triangle), create geometric constructions (e.g., see Fig. 28.1 for a worksheet description of one construction for the Triangles disk), and make measurements of lengths, angles, areas, and so on, of the drawings that result from the constructions. (*Note:* To make this same construction with the preSupposer, which doesn't have a built-in median option, label the midpoints of the three sides using the subdivide-segment option and then connect the vertex and midpoints using the segment command in the draw menu.)

The programs also store a record of users' activities as a procedure that can then be repeated on a new "initial" shape (e.g., see Fig. 28.2 for the construction from Fig. 28.1 repeated on an acute scalene, a right scalene, and an obtuse isosceles triangle). This repeat feature allows users to test the generality of their conclusions about the results of a particular construction (see Yerushalmy and Chazan [1990] for a more detailed description of the software and the ways it supports students in using diagrams).

With paper and pencil, one can do all that these programs do, but not as quickly or as accurately. This difference in speed and accuracy makes feasible an approach to the teaching of Euclidean geometry only theoretically possible with pencil and paper (for a more complete description see Chazan and Houde [1989]). With this approach, students' exploration becomes an important part of the course. Classes no longer meet only for the teacher's presentations to the whole group or for review of homework problems. Teachers pose inquiry problems to students, problems that are open-ended and lead to fruitful exploration (Yerushalmy, Chazan, and Gordon 1988). These problems, in contrast to traditional short exercises, are usually explored for one or more classroom periods, written about for homework, and then discussed.

---

Steps to make the drawing:

1. Start with an actute (scalene) triangle.
2. Draw a median in triangle *ABC* from vertex *A* (this step creates segment *AD*).
3. Draw a median in triangle *DAB* from vertex *D* (this step creates segment *DE*).
4. Draw a median in triangle *DAC* from vertex *D* (this step creates segment *DF*).

---

FIG. 28.1.

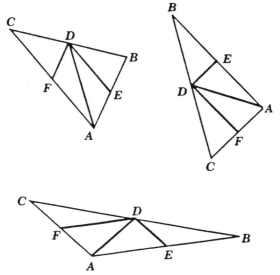

FIG. 28.2.   Scalene triangles with selected segments to midpoints

With the aid of the Supposers, students explore these problems, usually in pairs in a computer laboratory, and generate conjectures. Usually some of students' conjectures are true and some are false. Some of the false conjectures are easily modified to be true; others require major reworking. In class discussions, students share their conjectures and present arguments to support their ideas. As students are introduced to mathematical proofs and their facility with deductive proof develops, they are expected to present formal or informal deductive arguments for their statements.

One of the most important innovative aspects of this approach is that students are not trying to prove statements that they know are true (by virtue of being in the textbook) and that they know have been proved year after year in geometry classes. Some of the statements that they try to prove may not be true. Others that are true may not be present in their textbooks and may even be unfamiliar to their teachers (Kidder [1985] describes a student's conjecture that was unfamiliar to any of the mathematicians contacted by the author).

## INQUIRY SKILLS

Using this approach facilitates new goals for students and new standards for students' performance. Students should become competent explorers of open-ended problems. This new goal requires that students know how to work together and break down a large task into smaller tasks, generate hypotheses, use the computer to get feedback about their hypotheses, formalize their conjectures, generalize their conjectures, change and extend a problem, criticize or evaluate other students' conjectures, and develop arguments to support their own conjectures.

The rest of this article presents six important types of inquiry skills — verifying, conjecturing, generalizing, communicating, proving, and making connections — that students exploring a problem with the Supposers in a laboratory setting need to develop to be considered competent explorers of inquiry problems. These six skills present a starting point for describing the kinds of inquiry skills that students should develop. The approach illustrates how software like the Geometric Supposers can help us reach some of the goals set by the *Standards*.

Although students who possess these six skills will not necessarily be expert inquirers, in the author's experience they are important skills that teachers can help students develop. The order of presentation of these skills is pedagogically based, it is not the order of their use during a problem-solving session. Teachers may first want to focus on verification and then on some of the other skills, depending on the learning levels of their students.

## Verifying

For students to become nimble conjecturers, one of the most important prerequisite skills is to be able to verify, in specific cases, whether a geometrical idea seems to be true. Students need to know what data to collect and what measurements to make. To help students learn to verify statements, early in a high school or middle school course students can be asked to use the Geometric Supposer to make the construction described in Fig. 28.1 and then do the worksheet in Fig. 28.3.

Notice that in this problem students are supplied with statements; no pressure is placed on them to produce statements. To decide whether a particular statement is true, students must use geometrical definitions. Thus, in statement 1 students must know the definition of congruent triangles; in statement 2, the definition of a rectangle; and in statement 5, the definition of a trapezoid.

As students become more sophisticated geometers, they can be asked to do the same kind of task but with the minimum necessary measurements. Thus, for example, after learning SSS, SAS, and ASA, only three measurements are necessary to verify statement 1. Only one measurement is necessary to find a counterexample to statement 2. Students may also realize that in some circumstances measurement is unnecessary; that is, they may begin to see the relationships between different statements and make

---

• Are the statements given below true for your diagram (a diagram that results from the construction described in fig. 1)?

• Use measurements to support or contradict these statements. Record your decisions and the measurements on which they are based.

1. Triangle *BED* is congruent to triangle *DFC*.

2. Quadrilateral *AFDE* is a rectangle.

3. Segment *ED* is half as long as segment *AC*.

4. The line containing segment *FD* is parallel to the line containing segment *AB*.

5. Quadrilateral *AFDB* is a trapezoid.

6. Triangles *ADB* and *ADC* have the same area.

---

FIG. 28.3.

simple arguments based on definitions. Thus, if one of the statements had been that segments *AE* and *EB* have the same length, then students might come to understand, on the basis of the definition of a median, that this statement is always true by definition for the given construction. Or if in statement 4 students decide that the lines are parallel, they might realize that one does not need to measure to decide that *AFDB* is a trapezoid.

Teachers can also use this type of activity to —

• stress that one counterexample negates a "for all . . ." statement;
• have students start developing systems for collecting data; and
• ask students to make their own charts and learn to mark measurements directly on drawings.

## Conjecturing

To move from verifying to conjecturing, students need to understand that they can make their own statements and that most of the statements examined in school geometry are general statements — statements true for all members of some class. To help students learn to conjecture, teachers can add the instructions in Fig. 28.4 to the worksheet in Fig. 28.3. These instructions make explicit for students the fact that statements about a subset of all triangles will be valued. They also emphasize that most

> Make measurements to explore the relationships among the figures in your drawing. Can you make other true statements about your drawing? Write these statements and your evidence to support them.
>
> Next see if your statements are generally true. Use the repeat option to make new cases.
>
> 1. Write as many statements as you can that will always be true for drawings that are made by following the steps given in this problem. (The steps are listed in fig. 28.1.)
>
> 2. Write as many statements as you can that will always be true for drawings that are made by following the steps given in this problem *provided* that you start with a particular kind of triangle (e.g., isosceles or right triangles).
>
> 3. While you are working, record any geometrical questions or problems that you are wondering about.

FIG. 28.4.

conjectures are general statements about all members of some set of objects.

These instructions do not help students generate new conjectures or give them criteria for evaluating the value of their conjectures, important conjecturing skills that can be developed in other ways. If students working in the laboratory can't find any conjectures and have no ideas, then you can offer guidance by emphasizing that many conjectures are about relationships among geometrical objects. Teachers can ask students to identify the geometrical objects in a drawing created by the construction described in Fig. 28.1 and see whether they can discover a relationship between these objects. When deciding what sorts of relationships exist, teachers can remind students to check for the types of central geometrical relationships discussed in class: congruence, similarity, parallelism, and so on. When looking for numerical relationships, teachers can encourage students to look for inequalities, in addition to equalities, and for patterns involving ratios.

Another effective strategy to help students create conjectures is to suggest that students add additional segments to their construction. As new segments are drawn, new geometrical objects are created and new possibilities arise for relationships among geometrical objects.

Assessing conjectures is very difficult. The criteria of originality, "distance" from the original problem formulation (e.g., a conjecture about central points in polygons has evolved considerably from a problem posed about the three medians in a triangle), level of generality, and correctness are important factors (Yerushalmy 1986). Here are some key questions for students to ask themselves: "Are my conjectures supported by my evidence? Would other tests be worth performing? Have my conjectures been shown before? Are they a direct consequence of a known relationship? Is a more general idea or more specific idea represented here?"

## Generalizing and Posing New Problems

Some specific strategies can be used to help students examine the generality of their conjectures and pose new problems. One such strategy is to ask them to think about the set of objects for which their conjecture holds; in geometry, many times this set is the type of polygon being examined (in the construction described in Fig. 28.1, triangles). The instructions given in Fig. 28.4 encourage students to make conjectures about all triangles or the traditional subclasses of triangles.

Another useful technique is Brown and Walter's (1983) "what if not" strategy in which students systematically vary key aspects of a mathematical situation. In geometry, students can look for a numerical aspect of a construction. For example, the construction in Fig. 28.1 is carried out on triangles. You can encourage students to ask the following questions: "How should our construction be generalized for quadrilaterals? What might be a good definition of a median for a quadrilateral? Which aspects of the definition for triangles should be preserved? Which should be dropped?"

Another aspect that could be varied is the line segment used in the triangles. The construction in figure 28.1 uses medians. A median divides a segment into two congruent segments. Do important ideas to explore result if, instead of medians, one chooses segments that divide the opposite side of the triangle into three congruent segments? Do important ideas to explore arise if the medians are replaced by altitudes, angle bisectors, or other segments from a vertex to any point on the opposite side of a triangle?

In many circumstances, teachers find it difficult to allow students this kind of freedom of exploration. Yet, once students have developed a range of inquiry skills, this kind of activity can be extremely rewarding. By drawing auxiliary lines or by systematically varying some aspect of a problem to create new lines of inquiry, some students have developed very good problems. Indeed the construction in the problem in this article was one student's method for dividing a triangle into four triangular parts of equal area. Students then developed conjectures about this construction as an extension of that activity.

## Communicating

So far the description given here of verifying, conjecturing, and generalizing skills seems to describe a single student working alone. Yet students
rarely work alone with the Supposers; usually they work in pairs as they
explore constructions with the Supposers. When students work in pairs, it
is important to insist that they collaborate, that they share their thoughts.
Some students require help to develop the skills necessary to work well
together. They need to learn to listen to each other, to critique constructively and gently, to accept criticism, and to accept different working styles.
When students work together as a whole group to explore a problem or
when they discuss the results of an exploration, they must learn to present
their ideas to a group and to be patient as the class explores someone else's
ideas. Attainment of these communication skills is an indication of the
qualities of courage, honesty, and modesty that Lampert (1988) emphasizes
in Lakatos's (1976) and Pólya's (1954) descriptions of good mathematical
practice.

Beyond the work in pairs and with the whole class, students must
communicate their ideas in writing to their teachers; students working with
the Supposers write laboratory reports, individually or in pairs, summa-
rizing their explorations. These reports describe the conjectures they have
developed and present supporting arguments or measurements. In these
reports, students are expected to communicate their ideas clearly and in a
mathematically acceptable way. Conjectures must be intelligible to the
reader. Students must indicate what constructions they have performed, the
relationship that they conjecture to be true, and the set of objects for which
they think the conjecture is valid (Chazan and Houde 1989, 3). If students
give supporting evidence, the reader needs to be able to understand which
statement the evidence supports and why it is considered to be supporting
evidence. Once the teacher has introduced deductive reasoning, these
reports should include informal or formal deductive arguments as well.

## Proving

Once students are used to instructions like those in figure 28.4, the next stage
is to ask them to furnish supporting arguments for their conjectures. After
students have learned about proof, these arguments may be formal proofs;
yet the request for a supporting argument creates difficult questions for
students.

When students are working with measurement evidence to support their
conjectures, they have to decide when they have collected enough evidence
and when a statement requires a proof. For students, this decision involves
a question of allocation of resources: "Should I try to prove it now, but I am

not so sure it is always true; maybe I will be wasting my time?" versus "Should I continue to gather evidence, but I am pretty sure it is always true; maybe I will be wasting my time?" These questions are difficult ones and have no general answers.

Once students have decided to prove a conjecture, they must determine, in traditional parlance, "the given" and the "to prove." With the Supposer, the givens are to be found in the construction procedure; the "to prove" must be extracted from the wording of the conjecture.

Here are a few strategies for proving that are worth recommending to students:

- When working with a series of conjectures, put them in a sequence so that when the first is proved the second will be true, and so on. In this way, if you prove the first one, you will have proved many conjectures at the same time. If that is not possible, you can at least prove that the truth of the other conjectures is contingent on the truth of the first one.
- Write informal proofs by marking up drawings with different colors. For example, use pen for the givens and then mark the consequences of the derived steps in pencil.
- If you are unsure whether a step (not a reason) in your proof is true, check it with the Supposer's measurements.
- Consider drawing auxiliary lines; they make a proof easier to do by creating new objects and relationships between these new objects and the preexisting ones.

## Making Connections

It is also important that students make connections among different topics covered in class. The instructions in figure 28.4 ask students to write as many conjectures as they can. They do not suggest that students limit their conjectures to current classroom topics. Many times students will make conjectures that relate to material that was discussed previously or to material that has not been introduced.

If students have already studied congruence, when exploring the construction described in figure 28.1 they may connect congruence and equal areas. The construction produces a diagram with seven triangles (see Fig. 28.2). In nonisosceles triangles, triangles *DCF, DFA, DEA,* and *DBE* all have equal area; triangles *DCF* and *BDE* are congruent; triangles *DFA* and *AED* are congruent; and triangles *DBA* and *DCA* have equal area but are not congruent. Understanding that triangles with equal areas are not necessarily congruent is an important idea but one that many students find difficult.

If students have studied medians and know that a median in a triangle divides the triangle into two triangles of equal areas, then they may construct a proof for this theorem that is different from the one usually given. Instead of adding an altitude as an auxiliary line to prove that triangles $ADB$ and $ADC$ have equal areas, this alternative proof considers the two medians from vertex $D$ as auxiliary lines. Triangle $ADB$ is made from two triangles that are congruent to the two triangles that make up triangle $ADC$. Therefore, triangles $ADB$ and $ADC$ have the same area. These triangles are not congruent because the smaller congruent triangles are pasted together differently to make triangles $ADB$ and $ADC$.

## CONCLUSION

In the hands of dedicated and talented teachers, microcomputer tool software programs like the Geometric Supposer can help realize the central goals outlined in the *Standards*. Students exploring inquiry problems whose teachers work with them to develop the skills described in this article experience mathematics as problem solving, communication, and reasoning. For such students, the course is no longer fragmented into separate, unconnected sections; their explorations help them appreciate mathematical connections and structure.

Yet those of us who have been using the Supposer for the last four or five years are well aware that we have only begun to exploit the potential of this kind of tool. Tool software continues to grow and develop, partially in response to developments in hardware. (New versions of the Supposers are in development. Other packages for Euclidean geometry designed to make use of the Macintosh interface have also been developed [Cabri Geometre and The Geometry Sketchpad]. Graphing packages and Supposer-like instructional tools for algebra are also available.) As we acquire teaching experience with many examples of tool software, we will develop a wider variety of ways to use this technology effectively and a more sophisticated understanding of the ways in which it can support students' exploration in the classroom.

## REFERENCES

Brown, S. I., & Walter, M. I. (1983). *The art of problem posing*. Philadelphia: The Franklin Institute Press.

Chazan, D., & Houde, R. (1989). *How to use conjecturing and microcomputers to teach high school geometry*. Reston, VA: National Council of Teachers of Mathematics.

Kidder, R. M. (1985). How high schooler discovered new math theorem. *Christian Science Monitor*, April 19, 1985.

Lakatos, I. (1976). *Proofs and Refutations.* Cambridge: Cambridge University Press.
Lampert, M. (1988). The teacher's role in reinventing the meaning of mathematical knowing in the classroom. In M. Behr, C. Lacampagne, & M. W. Wheeler (Eds.), *Proceedings of the tenth annual meeting of the north American chapter of the international group for the psychology of mathematics education* (pp. 433–480). Dekalb, IL: Northern Illinois University.
National Council of Teachers of Mathematics. (1989). *Curriculum and evaluation standards for school mathematics.* Reston VA: Author.
Polya, G. (1954). *Mathematics and plausible reasoning (Vol. 1 & 2).* Princeton, NJ: Princeton University Press.
Schwartz, J. L. (1989). Intellectual mirrors: A step in the direction of making schools knowledge making places. *Harvard Educational Review, 59,* 51–60.
Schwartz, J. L., & Yerushalmy, M. (1985). The geometric supposers: Computer software. Pleasantville, NY: Sunburst Communications.
Yerushalmy, M. (1986). *Induction and generalization: An experiment in teaching and learning high school geometry.* Unpublished doctoral dissertation, Harvard Graduate School of Education.
Yerushalmy, M., & Chazan, D. (1990). Overcoming visual obstacles with the aid of the geometric supposer. *Educational Studies in Mathematics, 21,* 199–219.
Yerushalmy, M., Chazan, D., & Gordon, M. (1988). *Posing problems: One aspect of bringing inquiry into classrooms.* (Report No. 88-21). Cambridge, MA: Educational Technology Center of Harvard Graduate School of Education.

# Something Comes From Nothing Editors' Comments

Nothing comes from nothing;
Nothing ever could.
So somewhere in my youth or childhood,
I must have done something good.

The title of this section is inspired by words of the above song from the movie version of **The Sound of Music.** Walter's piece, **Generating Problem Posing From Almost Anything,** is excerpted and edited from a talk she gave that was subsequently published in two issues of a journal. With a renewed interest in what is called a constructivist view of learning, educators are becoming increasingly aware of the value of concrete materials and of playful behavior in the classroom. Where does one look, especially if one is interested in supplementing a text book approach to learning in the classroom?

One possible answer is found in the use of computers as described so well by Chazan in the previous section. During times of austerity, however, the computer is a tool that may be inaccessible to many school districts. Furthermore, the computer does not generate its own clever approaches to education. Where does one look?

The piece by Walter points to an array of interesting categories from which one can create mathematical activities—ones that are low cost. Some of what she talks about derives from the visual experience. You might wonder, as you read this article, what patterns you see every day but erroneously discard as not having mathematical worth. Among other things she shows how the activity of cutting and folding scraps made from a square has the potential to raise some questions that are rarely explored in standard text books. Simple diagrams—like a circle embedded in another circle—can be used to encourage youngsters to raise their own questions. She shows how children's games have significant mathematical content.

As you read the article, notice how some of the questions she raises in these everyday contexts derive from or expand upon the previously mentioned *Handy List of Questions* in **The Art of Problem Posing.**

# 29 Generating Problems From Almost Anything

Marion Walter

The theme of this conference is *economy*. There are quite a few ways of coping with *economy* if we had no text and no materials.

## SCRAP MATERIAL

Let us look at this plastic sheet which came from the inside of a chocolate box. You can see a grid on it where the chocolates were placed. There are lots of problems there. For example, the 4 rows on this sheet have 3, 4, 3, 4 cups in them, but the diagonals have 2, 4, 4, 3, 1 respectively (counting from right to left). Suppose there were 5 rows or 6, then how many cups would there be in each diagonal?

Suppose the numbers from 1 to 100 were written in such an array. In which row would 79 occur? There are many ways you can use such scrap material.

Let us remind ourselves that we have no bought materials and that we have to make do with what we have. Suppose that we do happen to have only one lonely square. Well, we can cut it in half and get two isosceles right triangles.

We can cut one of the triangles in half to get two smaller isosceles right triangles and we can do it again and again.

How many times can we do it before we can't pick up the pieces any more?

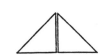

In theory everyone here can have a right isosceles triangle — and they are all similar to each other! What other 2-D shape(s), if any, can you find that you can cut in half to obtain two figures each similar to the original one? Can you find a 3-D shape that can be cut this way? It is no good having such triangles if we don't have a problem! If we put the two congruent triangles on top of each other; one high school level problem is to find the area of the overlap (see Walter, 1976).

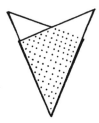

Try telling someone on the phone how to place the triangles as they are shown in the diagram! Young children can make patterns and designs with the various sized triangles and even work with some special fractions.

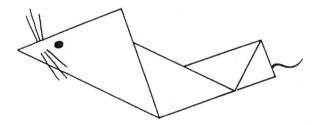

Suppose you don't even have a square; you have to make your own. In how many different ways can you make a square if you draw one? Fold one? Construct one with straight edge and compass? So many books show only one way and then they *tell* the students how to do it. See Walter (1981) for different ways to make a regular hexagon.

One can ask younger children to fold squares in half two different ways as shown and let them hold and touch these two different looking half squares. The two halves certainly look different; why should children believe right away that they have the same area?

Encourage children to fold quarter squares in different ways and to feel the quarters that look different. They can then unfold the squares and cut up the quarters to convince themselves that each quarter has the same amount. What about the rectangular paper mats we have at lunch that are thrown away after each meal? Try folding diagonal(s) to see halves and quarters. How should a young child know that the 4 parts are equal in area? It is a nice exercise to show, by cutting, rearranging, that the 4 triangles do indeed have equal areas.

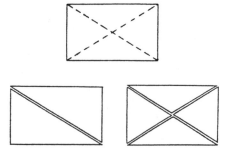

One problem that suggests itself when one folds a diagonal for the rectangle and then folds the flaps that cry out to be folded is to calculate the fraction of the rectangle that a triangular flap represents. Then fold the rectangle again so that the diagonal that was creased before, folds onto itself. The flaps again turn under which comes as a bit of a surprise though 'obvious'.

What is the shape in the middle — are you sure? Can you express the area of this rhombus in terms of the length l and width w of the rectangle? (see Walter, 1980).

## Doing Problems in More than One Way

Doing problems in more than one way is getting more for less, so to speak.

One problem which has appeared in many different places is to find the area of the petals in the diagram:

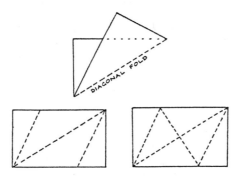

Using the circular paper coasters we have at lunch, we can cut semicircles and lay them out in a square to see the overlap on the overhead projector. It is a problem that can be solved in many different ways and changed in many ways. The diagrams show some of these ways.

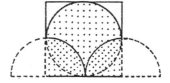

How else can the problem be changed? There are many ways. For now, note that we began with a square on the outside and 4 semi-circles inside. You get a different picture if you put a circle on the outside and 4 half squares on the inside.

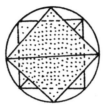

Are there other ways of doing this? We can pose the problem: what is the area of overlap? Can you solve it in several different ways? But can we also think of some new questions to pose? Note that here we turned the diagram inside out in a sense – and one way of changing a problem is to interchange parts! For younger children it is a problem just to analyze and copy such designs. Looking at these pictures makes me realize that we can get lots of problems from pictures.

## Problems from Pictures

One way to economise is to give children not a problem, but just the diagrams and have them make up their own problems from the diagrams. For example, if the picture is

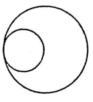

what problems might they or we pose? The picture here is simple, it could not be much duller but we can still pose some problems. Can we suggest some now to get started?

- Which is the bigger circle?
- How much do you have to roll the inner one to get back to the starting place?
- How many small circles fit into the large one?
- What is the area of the crescent?

- What is the circumference of each circle?
- What is the ratio of the area of the small one to the area of the large one?
- How big does the small circle have to be so that the ratio of their areas is say 1:3?
- If you draw an arrow on the tangent of the small one before you start rolling it where will it point as you get back to the starting point?
- What if the picture were of a three dimensional object?

Can you see how narrow some of the textbook questions are that ask just one thing? Then there is usually little thinking—students check the answers in the back of the book and that is the end of it.

If we put another picture beside the first one which is related to it—same size circles just in a different relationship, you will think of different problems. Even the same ones can excite new interest.

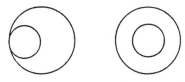

Can you think of some?

- What if it is a picture of a three dimensional object?
- In how many different ways can you place two circles?
- Investigate the picture as the circle moves to the right and as one participant said—do you want your egg fried or scrambled?

Here is one of my favourite bad examples that goes with this kind of picture.

If 200 years from now the textbooks are found, people will think chairs were always arranged behind each other. What is *the* question that goes with this picture in most elementary textbooks? 3 + 2 = ? What are some other questions? Can you make up some?

There was an enormous response to the chair challenge which resulted in many rich and interesting questions being compiled for use at all levels. We are able to include a very small selection chosen to illustrate this diversity. Why is it that textbook questions, on whatever level, are so uninspiring in comparison?

- *How many legs?*
- *How many different positions for the gap?*
- *Helen is sitting on the first chair, Edward is sitting on the last chair. How many moves must Helen make to reach Edward?*
- *If you had to stack chairs in twos, how many stacks could you make? What about stacking in threes?*
- *We need leg room to sit comfortably. What is the total distance from front of first chair plus leg room to back of last chair?*
- *Some are back to front. How many similar arrangements can be made?*
- *Five people sit on the chairs. In how many ways can they be seated so that they do not sit next to the same person twice?*
- *Five people on chairs, tallest at front, smallest at back. How many changes are needed to reorganise so that smallest is at front, tallest at back?*
- *Turn the chairs so that they all face the other way — but you have to turn 2,3,4, . . . at a time. What is possible? What not?*

## Problems from Making Up Problems

If one makes up problems or we ask our students to make up problems, we will find that we are often faced with new problems. For example, suppose that I am making up a simple geometry problem in which I want to have a parallelogram ABCD which is not a rectangle, where BE is perpendicular to AC.

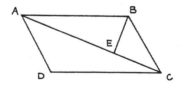

Now suppose we want to give as few angles as possible in order for students to find all the angles. How many angles do we have to provide? Which angles can they be? Is there any size of angle we cannot choose in order not to contradict the information? Might it not be useful for students to make up such a problem? I often ask my college students to make up problems for their class mates. There is a problem which can be found in several textbooks. It is one in which one must be careful in choosing numbers. Here is an example of this type of problem.

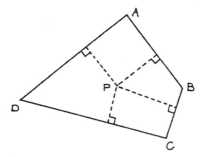

Line segments are drawn from a point inside a quadrilateral perpendicular to the four sides. The length of these segments and the length of the sides are given and the problem is to find the area of the quadrilateral. It is alas, easy to choose numbers that are impossible. As one of our students found out, she could not actually construct a quadrilateral from the given information. So, how can one, without just measuring, decide which numbers are legal ones for such a problem? Suppose you are given the length of the four sides, what can you decide about the length of the four segments for a given region for point P? I have not worked on the problem.

## Problems from Situations

We already saw that we can be led to a problem from the reseating we did earlier. Here is another problem that came up while flying from London to Seattle. This is what happened. We left Seattle at 9pm and had flown 3 hours so my unchanged watch said 12:00am. Someone looked out and saw quite a bit of light in the sky. Was it the Northern Lights, they wondered, or was it the moon, or could it be dawn breaking? That led to the question: What time was it now below us? You might need to know that there is an 8 hour time difference between Seattle and London and that it takes 9 hours to fly there. We might want to simplify the problem and assume that London and Seattle are on the same latitude and that we flew along a latitude rather than the polar route. What shall we do about the time zones? All kinds of questions arose among the passengers including ones about whether the earth was turning *and* the plane with it or the earth only but not the plane!

# MATHS FROM VISUALISING

If there is great economy we won't have many pictures except the ones in our heads (and new ones we paint)[1]
Close your eyes and visualise a horizontal line in front of you line and let a ball slowly roll along it.[2] I don't think the tape says that you should let it drop off but you might as well let it drop off as it will bounce right back again. Now would you fold up one section of your line so that when the ball rolls toward that end it can't fall off. Now roll the ball in the opposite direction and fold up that end of the line so that it can't roll off. How does your picture look now? If your picture looks a bit like an open box would you raise your hand? (Almost everyone does.) If your picture made a triangle would you raise your hand? (About 4 or 5 people did.)

That raises a lovely problem. If you have a line, and let's simplify the problem and say you can bend it only at unit intervals, which bends will give you triangles? How many different triangles can you form this way? (Later you might want to ask what the probability is of obtaining a triangle if you make two bends at random). Let's take an example.

Suppose you have a 10 cm stick which can bend only at cm intervals, how many different triangles can you obtain? Young children can solve the problem by using straws that they mark at cm intervals. Even older students sometimes are powerfully reminded of the triangle inequality which they sometimes first ignore in counting the 'triangles'. It is perhaps surprising that one can get only two triangles and that they are both isosceles. If you work with different lengths to obtain data—it is not very easy to see what is going on! If enough data is generated older students may be able to notice that the number of triangles of length n (n even) equals the number of triangles for length (n − 3); that is T(n) = T(n − 3) for n even and n ≥ 6 where T(n) stands for the number of non-congruent triangles that one can form with integral sides if a length of n is provided. The general formula is advanced math (see Dearborn, 1986; Jordan, Walch, & Wisner, 1979).

Let's look at one more of the visual exercises on the *Imaginings* tape and see how we can use it as a starting point for more problems. *Close your eyes* again and visualise a triangle—vertically in front of you. You might want to make it acute or obtuse—or skinny—though skinny is not a mathematical word. Eventually make it equilateral. Imagine it white. Now imagine a small equilateral triangle in each corner—imagine these black. The tape then asks what shape is the white piece? Could you let your small triangles

---

[1]ATM (1982) and the DIME Materials (not dated) are particularly rich in work requiring visualization.
[2]The beginning activity comes from an audio tape entitled *Imaginings* and was produced by the British group known as Leapfrogs available originally from Tarquin publishers, but is no longer available.

grow and grow so that the resulting figure in the middle is a regular one—what regular figure does it become first? Can you make the black triangles grow a bit more so that the figure in the middle becomes a regular something else? What does it become? And if you grow the triangles more what happens then. Can we now use this visualising problem to make even more problems? Let's list some attributes of some of the pictures we saw. How could we describe it?

- it has symmetry
- it is a triangle
- it has an irregular hexagon inside
- the corner shapes are triangles
- the corner shapes are the same shape as the outside shape
- it is two dimensional

It is interesting to ask what if it were three dimensional? We would have a large regular tetrahedron with a small regular tetrahedron at each corner. If we cut off each of the small tetrahedra,

what would the remaining shape look like? What if we first grew the small tetrahedra until they touched and then cut them off? What would the shape inside look like? Can you picture it? One way of generating problems is to list some of the attributes of your starting point—we listed only a few—and to pick one or two of them and ask, 'What if it were not so?' For example, what if it were not two dimensional, what if it were not a triangle? One alternative is that it is a square and that there are still triangles in the corner—not equilateral but perhaps right isosceles. Or there could be squares in the corner. We could ask the same questions as we asked before and we could again try visualising the shape in the middle and see what happens as the corner shapes grow. Of course you will also think of many other new questions suggested by the pictures (see Brown & Walter, 1983).

## Scrap and Other Available Material—Again

When I tried to think of what material is either often thrown away or is easily available, several things came to mind. First, we always have some water available. What kind of problems can we have with water? Well, there are the well known Piaget type problems and experiments: which container holds the most, the tall skinny one or the short fat one? But there are also the supermarket deception kind of problems. One nice simple exercise you can do on your way out from this lecture hall. Look at the different bottles we have on the table and order them according to size just by looking at them.

I know that I have been fooled in a supermarket by 12 oz jam jars that looked like one pound and 13 oz 'pound' coffee tins. These containers are very cleverly designed to try and fool us. You might want to work on and explore these deceptive packages that look as if they hold more than they actually do and design packages that deceive so that you and your students become more aware of the deceptions practised.

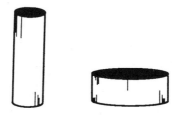

*Which holds the most, the tall skinny one or the short fat one?*

There are many kinds of cardboard tubes – kitchen paper towel ones, toilet paper rolls and xerox paper rolls which are very sturdy indeed. What can you do with paper rolls? You can print circles with them – useful for number chart work – circle the multiples of 3 etc. A sophisticated problem is to examine how they are made. Young children could just unroll them. During the first workshop I did with materials in Lancaster in 1971 we made plaster casts from them. You get the helix mark on the cast which is nice. You can also cut the rolls and rejoin them. If you don't cut perpendicular to the roll, the cross section is not a circle and you can join the two pieces in only two ways. You can make some nice shapes this way – and raise some nice problems. Can you get the construction to close again for example?

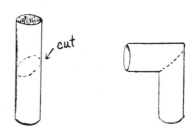

What else do we throw away all the time? Milk cartons! There are many things you can do with them. Let's only look at the cartons that we have cut down to form cubes without a top. How many squares are they made of? One of the things you can ask is, 'How many ways are there of arranging 5 squares?' Here are just a few:

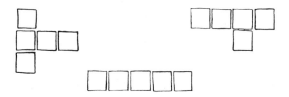

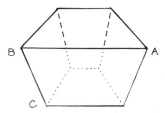

Some of them fold into open cubes and some don't. How many different patterns are there and how many of them fold into boxes without tops? Which patterns fold? Number the ones that do 1, 2, 3, . . . Now write one of these numbers on the bottom of your milk carton and try to cut it so as to obtain the pattern whose number you wrote on your carton (see Walter, 1968, 1969, 1971).

Here is another box with all sides congruent trapezoids.

It raises all kinds of questions about the size of possible angles. What is the smallest that angle ABC can be? the largest? Is there a smallest? A largest? What happens in the extreme case? What is the volume? How much paper is used? Is it more or less economical to make than a cube?

Then there is much work to be done with cereal type boxes. Collect some of them and examine them for volume and surface area. You will be surprised! Also you will find some pairs of containers where the one with larger volume actually contains fewer cornflakes! Why do manufacturers make narrow containers that waste material is a question worth discussing with your students. You can also estimate the amount of material saved if say 10,000 boxes of a given volume were made with better dimensions. There is a lot of ordinary curriculum maths in this type of investigation.

There are different problems that come up when you use scrap material. Some questions come to mind *before* one uses the material, some *while*

using the material and some *after* the material has been used. You might
also want to look at the curriculum ideas involved.

## Games

As a child I used to run up and down stairs. You can make up many
different problems with stairs! Here is one: You can go up a certain
staircase 2 at a time and have 1 stair left and go down 3 at a time and have
1 stair left. Does this determine the number of stairs? What if you are told
that I can jump down 5 at a time and then there are none left over? Does
that determine the number of stairs? If not, what else could you be told that
would? You can make up a lot of different problems while you jump up and
down stairs!

Noughts and crosses, known to Americans as tic-tac-toe, can give rise to
a lot of problems:

• How many different starting points are there?
• What shall we decide is meant by different?
• What are some strategies to win?
• How can you record information without drawing pictures?
• What if you had a 4 by 4 grid?
• What if you played noughts and crosses in 3 dimensions?

Dominoes lend themselves to many problems. Children do a lot of counting
while skipping rope. There are problems about one child double jumping
every third beat and another every fifth beat. It was the potato race that
made me think a lot when I was young. Recall that you have to pick up the
potatoes, placed in a row, one by one and place them in a bucket before
running to the finish line. I used to really wonder whether it was better to
first pick up the one nearest or the one furthest away! Or would the middle
one be best? Then there was also psychology involved as you watched others
and saw how many potatoes were left! If I had 3 left—near me but the
person running next to me had only 1 left further away. . . .

You might want to alter the potato race. Let's list some attributes—5
potatoes, one at a time, all in one line, at equal distances . . . What if you
had, say, ten potatoes, not all one by one, not all at same distance from
each other—what kind of race can you make up?

And then there was the three legged race. If there are 24 'legs' running
how many people are in the race? If there are 48 people running how many
legs will be running?

I recall in the game of Tag where one could be safe by standing in front
of a pair of players and the one outside would be IT. I recall having a

distinct feeling for when something is added and the same amount is taken away that the result is the same.

Perhaps it is worthwhile to think back on the games you played and to consider what mathematical ideas may be lurking in them.

Card tricks of course too can be used to do a lot of maths. I have purposefully not mentioned games that have obvious maths in them.

### Extending the Problem — Don't Always Keep to Special Cases.

We spend a lot of time in the school curriculum on special topics without ever telling the students how special the cases are. For example, we spend a lot of time in geometry on congruent triangles. Do we bring up the question of why triangles may in fact be more important to study than quadrilaterals? Do we raise the question of what congruency conditions for quadrilaterals would be? (see Ranucci, 1973; Spitler, 1976). Perhaps my favourite example is one which I brought to the ATM some years ago. The problem arose when I was making plaster casts of geometric shapes with youngsters and we were making moulds for right circular cones. When the students had made some, I suggested they now try and make some moulds for a non right circular cone — an oblique cone with a circular base (not a cone obtained by slicing a right circular cone at an angle). They knew that they should draw the plan. Luckily the bell rang. I spent a lot of time trying to do it and then took the problem to the ATM meeting.

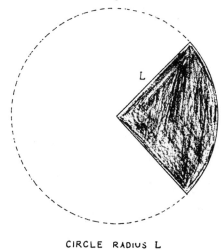

CIRCLE RADIUS L

*The shaded part is a plan of the right circular cone overpage.*

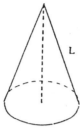
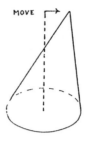

RIGHT CIRCULAR
CONE

A little change of distance for the vertex alters the plan drastically, Bill Brookes sent me a solution . . . It is not an elementary problem but it, too, is an example where we don't tell children how special a situation is. Surely we want children to wonder what happens if . . . or what happens if not . . .? You might be on the look out for other special situations.

Well there are many other situations from which you can get problems from real situations; for example when next you see stacked oranges you may ask yourself why are they stacked the way they are — how many on the next layer and so on. (see Ranucci, 1974; Sachs, 1974). And I won't have time to talk about Maths from people — Nuffield did a lot of that.

## REFERENCES

Association of Teachers of Mathematics. (1982). *Geometric Imaginings*. Author.

Brown, S. I., & Walter, M. I. (1983). *The art of problem posing*. Hillsdale, NJ: Lawrence Erlbaum Associates.

Dearborn, R. G. (1986). Integral sides. *Mathematics Teacher, 79,* 238.

DIME Materials (undated). Stradbroke, England: Tarquin.

Jordan, J. H., Walch, R., & Wisner, R. J. (1979). Triangles with integer sides. *American Mathematical Monthly, 86,* 686–689.

Ranucci, E. R. (1973). The congruency of quadrilaterals. *Mathematics Teaching, 64,* 35–37.

Ranucci, E. R. (1974). Fruitful mathematics. *Mathematics Teacher, 67,* 35–37.

Sachs, J. M. (1974). A comment on fruitful mathematics. *Mathematics Teacher, 67,* 701–703.

Spitler, G., & Weinstein, M. (1976). Congruence extended: A setting for activity in geometry. *Mathematics Teacher, 69,* 18–21.

Walter, M. (1968). Polyominoes, milk cartons and groups. *Mathematics Teaching, 43,* 12–19.

Walter, M. (1969). A second example of informal geometry: Milk cartons. *Arithmetic Teacher, 16*(5), 365–372.

Walter, M. (1971). *Boxes, squares and other things*. Reston, VA: National Council of Teacher of Mathematics.

Walter, M. (1976). Two problems from a triangle. *Mathematics Teaching, 76,* 38.

Walter, M. (1980). Mathematizing with a piece of paper. *Mathematics Teaching, 93,* 27–30.

Walter, M. (1981). Do we rob students of a chance to learn? *For the Learning of Mathematics, 1*(3), 16–18.

Walter, M. (1981). Exploring a rectangle problem. *Mathematics Magazine, 54,* 131–134.

Walter, M. (1985). The day all textbooks disappeared. *Mathematics Teaching, 112,* 8–11.

# Your Turn
# Editors' Comments

As in the previous chapter we end this section with an opportunity for you to explore some of the problem posing strategies of this chapter, this time in a geometric context. We select a physical object — one that is easy to create and one that we introduced in **The Art of Problem Posing,** a geoboard. Below is a picture of it:

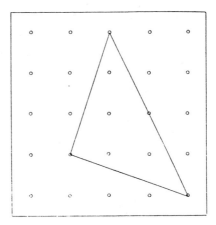

A standard geoboard is a square wooden board with 25 nails in it equally spaced as indicated. Accompanying a geoboard is a set of rubber bands that can be placed around the nails as indicated. If you are given one geoboard as above, use as many of the strategies as you can from articles in this chapter to pose problems on the board. Even if you are familiar with the board—perhaps having read about it in **The Art of Problem Posing**— you should now have added machinery to devise some interesting new problems to explore. In order to get started if you have not played with the board before, either create one on your own or make a number of copies out of graph paper that simulate the board and put real or imaginary (e.g. using a pencil on a graph paper that simulates a board) rubber bands around the nails. At first try to do so without carrying out any variations on the board itself. Pose problems about shapes that are created by the bands.

After you have played with the board a while, find someone else to accompany you in further exploration. After you have made up questions that "accept the given," see if you can create a number of them that ask What-If-Not. In order to do so, ask yourself what the attributes of the geoboard are; vary those attributes; ask questions about the new phenomenon and explore.

When you have done some of the above exploration (and you might wish to take a considerable period of time—perhaps even a couple of weeks), think about the following What-If-Not on the geoboard. Suppose you allow for the possibility that one of the little squares is considered *off bounds* (the nails surrounding the square remaining in place however). Now do some problem posing that accepts the new given, followed by a What-If-Not on the new given.

After doing some of the above activities, read the article by Schmidt entitled, **A Non-Simply Connected Geoboard—Based On the "What If Not" Idea.** Which of his explorations did you anticipate? Which new ones did you create? Give some thought to how it is that some of the exploration you conducted might fit into the existing curriculum. How might concepts such as area, perimeter, fractions, the Pythagorean Theorem and others be enriched by this experience?

# 30 A Non-Simply Connected Geoboard – Based on the "What If Not" Idea

Philip A. Schmidt

The idea for a non-simply connected geoboard occurred to me while reading "What If Not?" by Marion Walter and Stephen Brown (1969). Their purpose in writing "What If Not?" was "to present and analyse a method for generating new curriculum ideas." The authors recommended that in dealing with a particular phenomenon, one make a list of its attributes and then ask, "What if not?" In other words, ask how these attributes might be modified to produce other models of mathematical interest.

One of the attributes of the standard geoboard is its "simply connectedness," and a corresponding alternative is non-simply connectedness. (By a simply connected geoboard we mean a geoboard such that for each peg $P$ on the board, we can shrink the entire board to $P$.) The geoboard discussed here is non-simply connected. I shall show that most of the activities for which the standard geoboard is used can be performed on the modified geoboard. I shall exhibit tasks unique to a geoboard with a hole in it. Furthermore, a look at problems that are either different from the standard geoboard problems or that require a different analysis on the "holy geoboard" should help clarify other attributes of the geoboard and suggest new ideas.

## DESCRIPTION OF THE GEOBOARD

The geoboard we shall use is the standard device with five rows of nails and five columns of nails. Geoboard I will refer to such a standard geoboard (see Fig. 30.1). Geoboard II will refer to the same geoboard with square

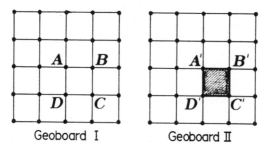

Geoboard I        Geoboard II        FIG. 30.1.

$A'B'C'D'$ punched out, but with the nails of the deleted square remaining and a rubber band permanently connecting the four vertices $A'$, $B'$, $C'$, and $D'$.


## SHAPES, AREAS, PERIMETERS

### Introductory Play

I found that, among students who had played with the standard geoboard, some had no trouble adjusting to geoboard II, but others found it extremely difficult. For example, some students had difficulty doing on geoboard II problems that gave them no trouble on geoboard I (i.e., problems that are not at all altered by posing them on geoboard II). It may be that students with this difficulty have perception problems when dealing with non-simply connectedness. (Although I have not done extensive work in correlating difficulty in handling board II with other perception problems, the reader might like to do so in order to determine the potential of geoboard II as a diagnostic tool.)

After the student has had time to get acquainted with both geoboards, ask the following questions:

1. How do the two geoboards differ? (Possible answers: "Geoboard I has area 16, but geoboard II has area 15"; One has a hole in it, but the other does not"; "There is no difference."

2. Geoboard I is a square. Is geoboard II a square as well?

3. Construct any region on geoboard I, first using one rubber band, then using more than one rubber band. For each region, construct its corresponding region on geoboard II. In each case, how do the corresponding regions differ, if at all?

4. Construct on geoboard I a region that does not include square $ABCD$, and construct its corresponding region on geoboard II (Fig. 30.2). (The student should be allowed to convince himself that if the region on

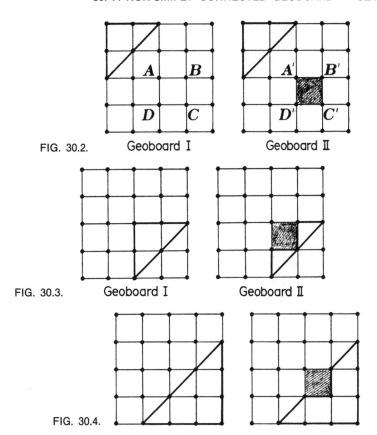

FIG. 30.2.    Geoboard I         Geoboard II

FIG. 30.3.    Geoboard I         Geoboard II

FIG. 30.4.

geoboard I does *not* contain *ABCD,* then the corresponding region on geoboard II will not differ; the two areas are therefore the same.)

5. Ask the student to construct on geoboard I a region that includes *all* of the interior of square *ABCD,* and have him calculate the area of that region. Then have the student construct the corresponding region on geoboard II and calculate the area of that region (Fig. 30.3).

6. Once the student has mastered the problem in question 5 above, a related problem can be offered to him. Construct on geoboard I a region that contains only a *part* of the interior of square *ABCD.* Have the student form the corresponding region on geoboard II and compute or estimate areas (Fig. 30.4).

## More Difficult Problems

When the student feels at ease with this new geoboard, many more problems can be posed. Here are some examples of questions that might be thought-provoking for the student:

1. When will the figure on geoboard I have the same area as its corresponding region on geoboard II? Can you find a way of predicting when the results will be the same and when the results will be different without actually looking at the boards?

2. When will the area of the region on geoboard II be one-half the area of the corresponding region on geoboard I? One-fourth? and so on.

3. Give an example of a region on geoboard I such that the corresponding region of geoboard II will be one unit smaller; a region such that the corresponding region will be one-half unit smaller.

4. Give examples of regions on geoboard I whose corresponding regions on geoboard II have area zero. Generalize.

5. Regions of area one can be constructed on both geoboards. Is there a number such that regions on geoboard I can have that area but regions on geoboard II cannot?

6. When will the number of sides of a region on geoboard II be the same as on geoboard I? When will this number be more? Less? (See Fig. 30.5.)

7. When will the perimeter of a region on geoboard II be the same as on geoboard I? More? Less? (See Fig. 30.6.)

8. How many squares (triangles) are there on geoboard I? On geoboard II?

9. When will the resulting region on geoboard II *not* be simply connected?

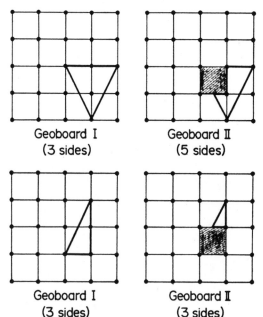

Geoboard I
(3 sides)

Geoboard II
(5 sides)

Geoboard I
(3 sides)

Geoboard II
(3 sides)

FIG. 30.5.

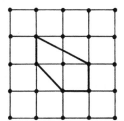 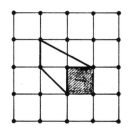

FIG. 30.6.  Perimeter the same
on both geoboards

10. It can be shown that the area $A$ of a region with $i$ nails on the inside and $b$ nails on the boundary is $A = \frac{1}{2}b + i - 1$ for geoboard I (this is Pick's formula). What is the formula for geoboard II? (*If* such a formula exists, find it. See Marshall [1970].)

## GAMES

### Geometric Clues

Geometric Clues is a geoboard game that has been used with success in the classroom. One person in the class constructs a region on his geoboard using one rubber band. No one else knows what his figure is. The other students ask questions of the person who constructed the figure and try to build an exact copy of the constructed figure on their own geoboards.

The game can be played on geoboard II and takes on an interesting twist if the person answering the questions is referring to geoboard II but everyone else must interpret his answers onto geoboard I.

### Taxicab Geometry

In this exercise, the geoboard is imagined to represent a city laid out in square blocks. There are no one-way streets. A taxi traveling from one point to another must never add unnecessary mileage. Assume all trips start from the upper left-hand corner of the geoboard. Assume also, for the purpose of this game, that $F'$, $G'$, $H'$, and $J'$ are completely out of bounds. (See Fig. 30.7.) How many different legal routes are there—

1. driving from $A$ to $B$? from $A'$ to $B'$?
2. driving from $A$ to $D$? from $A'$ to $D'$?
3. driving from $A$ to $C$? from $A'$ to $C'$?
4. driving from $A$ to $E$? from $A'$ to $E'$?

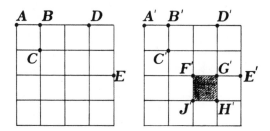

FIG. 30.7.

For which destinations does the missing square alter the result? By how much?

## Moving the Square(s)

So far the alternative corresponding to simply connectedness is to make one hole in a particular place on the geoboard. We can now, again, ask ourselves the question, "What if not?" This could lead us to other alternatives. Suppose that any square or combination of squares could be removed from the geoboard (see Fig. 30.8). What would the consequences be? What would be some new questions? All the previously asked questions (concerning area, perimeter, number of sides, etc.) can again be posed with regard to these new boards. In addition, Taxicab Geometry and Geometric Clues can still be played. The following new questions can also be posed:

1. Given a figure on geoboard I, what hole(s), if any, would you punch out to maximize the perimeter? To minimize the perimeter?

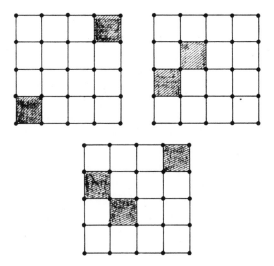

FIG. 30.8.

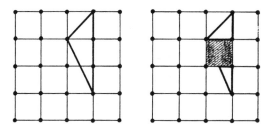

FIG. 30.9.

2. Given a figure on geoboard I, what hole(s), if any, would you punch out to maximize the number of sides? Minimize the number of sides? (See Fig. 30.9 for the effect of punching a particular hole out on the number of sides in a region.)

## SUMMARY

It is important to note at this time that I have tampered with only one attribute of the standard geoboard and have worked with only one (or two) of its alternatives. There are certainly many other possibilities. For example, suppose that we remove exactly one-half of every one-unit square of the ordinary geoboard (Fig. 30.10). The reader will, no doubt, be as surprised and motivated as I was by the number of rich ideas he gets by picking one attribute of the geoboard and asking himself the question, "What if not?"

I do not want to leave the reader with the impression that "what if not" is applicable solely to mathematics. Although my attempted reaction and application of the scheme is primarily a mathematical one, I am convinced that the "what if not" procedure is applicable to most, if not all, fields of inquiry. Moreover, as I hope I have conveyed in this article, the process is a remarkably interesting, effective, and useful one.

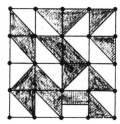

FIG. 30.10.

# REFERENCES

Brown, S. I., & Walter, M. I. (1970). What if not? An elaboration and second illustration. *Mathematics Teaching, 51,* 9-17.

Krause, E. F. (1973). Taxicab geometry. *Mathematics Teacher, 66,* 695-706.

Marshall, A. G. (1970). Pick with holes. *Mathematics Teaching, 50,* 67-68.

Walter, M. I., & Brown, S. I. (1969). What if not? *Mathematics Teaching, 46,* 38-45.

Walter, M. I., & Brown, S. I. (1971). Missing ingredients in teacher training: One remedy. *American Mathematical Monthly, 78,* 399-404.

# Author Index

327

# Subject Index

## A

Accepting the given, 4, 23-24, 205, 207, 279, 318

## B

Brute force problem posing, *see* Accepting the given

## C

Challenging the given, xiv, 4, 23, 124, 270
Computers, 160, 163, 184-185, 198, 200, 289, 298
  The Geometric Supposers, 289-292, 296-298
Conjecturing,134, 136-137,140, 182, 184, 190, 193-200, 280, 289, 291-297
Critical thinking, 62, 93-94
Curriculum and evaluation standards, xiii, 89-90, 134, 184, 235, 241, 289

## D

Dualistic view of mathematics, *see* Perry developmental scheme

## E

Editorial boards, 3, 7-9, 14-15, 25
Errors, *see* Mistakes

## G

Games, *see* Number tricks
Generalizing, 45, 49-50, 190, 192, 237-238, 240, 246, 279-280, 283, 287, 292, 295
Geoboards, 10, 44, 46, 180, 317-325

## H

Handshake problem, 74-76
Humanistic education, 61, 66, 68, 88, 91-98, 102, 187, 232-233, 249-250, 264-265, 273, 277-278
Hypothesizing, *see* Conjecturing

## J

Journals, 8-9, 11, 25

## M

Mathematics anxiety, 82-83, 187
Mathematization, 101

# Notes on Contributors

**Elizabeth D. Bjork** received her masters degree in mathematics education from Boston University. She has been a classroom teacher, supervisor, developer and trainer of teachers in mathematics. In addition, she has worked with both undergraduate and graduate students in mathematics, language arts and computers. Presently at Education Development Center in Newton, Massachusetts, she co-directs mathematics curriculum projects as well as the assessment of district mathematics programs. She also guides curriculum development and teacher education for the Elementary Mathematics Project.

**Rick N. Blake** teaches both graduate and undergraduate courses in mathematics education at the University of New Brunswick in Canada. His areas of interest include geometry and problem solving—in particular, working with teachers at the middle school level in developing geometric concepts and relationships through problem solving and problem posing.

**Raffaella Borasi,** a native of Italy, holds a Ph.D. in mathematics education from the University at Buffalo. She is currently Associate Professor of Education at the University at Rochester. Her recent book, *Learning Mathematics through Inquiry,* focuses on her interest in developing instructional implications of an inquiry approach to the field.

**Stephen I. Brown** holds a dual appointment as professor of mathematics education and philosophy of education at the University at Buffalo. He and Marion Walter have collaborated on numerous writing and teaching projects since they both were faculty members at Harvard Graduate School of Education in the late 1960s. In addition, he has published books and articles on epistemological and humanistic themes in education dealing with issues such as the roles of discovery, intuition, humor, surprise, doubt, progressive education and personhood in mathematics.

**Dan Brutlag** has taught mathematics at the secondary school and college levels for eighteen years. Currently he is the Project Coordinator and primary writer for a mathematics curriculum development project for grade eight, funded by the National Science Foundation. He is at the University of California in Oakland.

**Dorothy Buerk** received her Ph.D. in mathematics education from the University at Buffalo and is now on the faculty at Ithaca College in New York. She works with those who would prefer to avoid mathematics. She emphasizes writing, humanistic mathematics, metaphors for mathematics and cooperative learning. She gained insights into her students' fear of mathematics by reflecting on the way she overcame her own fear of deep water. She now is an avid swimmer and white water canoeist.

**William S. Bush** received his Ph.D. in mathematics education from the University of Georgia and is presently Assistant Professor of Mathematics Education at the University of Kentucky. His primary teaching responsibilities are methods courses in primary and secondary school mathematics.

**Charles Cassidy** has been a member of the Department of Mathematics and Statistics at Université Laval in Quebec since 1970 and presently holds the title of *Professeur Titulaire*. His areas of research are group theory, number theory, and most recently problem solving and teacher training at the primary and secondary school levels.

**Daniel Chazan** is a Dow-Corning clinical assistant professor at Michigan State University. In order to aid his research in mathematics teaching and learning, he teaches algebra at a local public high school. His interests include using computers to support student exploration and the potential of philosophy of mathematics to inform the teaching which emphasizes such exploration.

**Tommy Dreyfus** teaches mathematics to pre-service teachers of engineering at the Center for Technological Education in Holon, Israel. His research addresses cognitive aspects of learning mathematics, especially at the college level. He is particularly interested in the role of visualization and the use of computers.

**Werner Feibel** received his Ph.D. in psychology from the University of California at Santa Clara. After having taught and done research for several years in the field of psychology, he switched to computer science and helped develop and supervise the programming of computer-based learning materials. He now writes about programing and other computer related topics and does occasional teaching at local universities.

**Ann Fiala** is a reading specialist and teaches a fifth grade gifted and talented class that integrates language arts with mathematics. She teaches in the Victoria Independent School District in Texas. She collaborated with William Bush in the writing of the article in this collection when he was an assistant professor at the University of Houston – Victoria.

**David Fielker** has taught mathematics in secondary schools and has been the director of a mathematics centre in England since 1967. Former editor of the British journal *Mathematics Teaching* (from which a number of articles in this collection appear), he is presently interested in the reform of geometry in the secondary schools and in the integration of calculators in primary and secondary mathematics syllabi.

**Alex Friedlander** works as a curriculum developer and teacher educator at the Weismann Institute of Science in Rehobot, Israel and also teaches at the School for Natural Sciences in Tel Aviv. His current interests lie in intuitive and informal approaches to algebra and in the design of enrichment activities for junior high school students and teachers.

**E. Paul Goldenberg** is a Senior Scientist at Education Development Center in Newton, Massachusetts. Interested in the role of curriculum and software in mathematics learning, he is the author of *Special Technology for Special Children, Exploring Language with Logo,* and is editor of the MIT Press series *Exploring with Logo.* Formerly a mathematics coordinator for the Laboratory Schools of the University of Chicago, he has also taught in grades 2 through 12 as well as at college and graduate levels.

**Bernard R. Hodgson** received his Ph.D. from the Université de Montréal, and has been a member of the Department of Mathematics and Statistics at Université Laval in Quebec since 1975. He is now *Professeur Titulaire.* Besides his research interests in mathematical logic and theoretical computer science, his main professional activities concern teacher education both at the primary and secondary school level. He was one of the main organizers for the 1992 International Congress in Mathematics Education in Québec.

**Larry Hoehn** received his Ed.D in mathematics education from the University of Tennessee. He has taught both secondary school and college level mathematics. Since 1979, he has been on the faculty of the Department of Mathematics and Computer Science at Austin Pea State University in Clarksville, Tennessee. He has published widely and was the recipient of the faculty award for scholarship.

**Douglas L. Jones** received his Ph.D. in mathematics education from the University of Georgia and is an Assistant Professor of Mathematics Education at the University of Kentucky where he teaches mathematics methods classes and is involved in the Kentucky K-4 Mathematics Specialist Program. He conducts research on the kinds of changes teachers make in responding to educational reform. He continues to enjoy involving his methods students in problem posing.

**John R. Jungck** is the Director of BioQUEST, a national consortium for curricular innovation in biological education whose philosophy is grounded in problem posing, problem solving and persuading peers. He is the past editor of the *American Biology Teacher* and is the editor of *Bioscene: Journal of College Biology Teaching*.

**Barry V. Kissane** was a secondary school teacher in Western Australia before taking a position as a mathematics teacher educator. He teaches pre-service primary and secondary teachers of mathematics, as well as in-service and post graduate students of mathematics education at Murdoch University in Perth, Western Australia.

**Lawrence Meyerson** received his Ed.D. in mathematics education at the University at Buffalo and taught for several years at Northern Michigan University. He subsequently received his law degree from Seton Hall University and is presently a partner in the firm of Rubenstein, Rudolph, Meyerson & Billings in Oakland, New Jersey. While at law school he received the *Corpus Juris Secundum* Award for contributing the most significant legal scholarship to the class.

**Barbara M. Moses** teaches mathematics to pre-service and in-service teachers in the Department of Mathematics and Statistics atg Bowling Green State University. Her research focuses on the role of visualization in the problem solving process, as well as the use of technology to improve this spatial thinking.

**Israel Scheffler** is Victor S. Thomas Professor Emeritus of Education and Philosophy at Harvard University. He served on the Harvard faculty from 1952 to 1992. A Fellow of the American Academy of Arts and Sciences, a charter member of the National Academy of Education and a past president of the Philosophy of Science Association, he was co-founder and co-director of the Philosophy of Education Research Center at Harvard. He has published a number of books in the philosophy of education and philosophy of science, the most recent of which are *Inquiries* and *In Praise of the Cognitive Emotions*.

**Philip A. Schmidt** received his Ph.D. in Mathematics Education from Syracuse University. After chairing the Department of Mathematics at Berea College, he joined the faculty of the School of Education at the State University College of New York at New Paltz. Currently the Dean of that school, he is the author of numerous grants and journal articles and is co-author of *College Mathematics*.

**Kenneth Shaw** is Assistant Professor in Mathematics Education at Florida State University. His teaching focuses on helping prospective teachers better understand the dynamics and applications of mathematics. His research interests include learning about the impact of the culture and teachers' beliefs on teacher change.

**Marian Small** is Professor of Mathematics Education at the University of New Brunswick in Canada. She teaches primary and secondary school level mathematics curriculum and methodology courses and is currently co-authoring a K–6 text series which views mathematics as a set of ideas with which to interact rather than a set of skills to master.

**Larry Sowder** received his doctorate from the University of Wisconsin. Having taught high school mathematics in Indiana and having served on the faculty of Northern Illinois University, he is presently at San Diego State University and continues to be involved in pre-service and in-service teacher education.

**Marion I. Walter** established the mathematics major program at Simmons College and subsequently joined the Harvard Graduate School of Education, where she and Stephen Brown began their teaching and writing collaboration. In addition to problem posing, she is particularly interested in links between mathematics and the visual arts, and has written numerous books and articles dealing with intuitive notions of geometry. Since 1977 she has taught at the University of Oregon.

**David J. Whitin** is an Associate Professor of Elementary Education at the University of South Carolina. He has co-authored *Living and Learning Mathematics* — a text that focuses upon mathematics as a way of thinking. He has also co-authored *Read Any Good Math Lately? Children's Books for Mathematical Learning K-6* which uses children's literature as a springboard for mathematical explorations.